C0-AMZ-831

PERGAMON INTERNATIONAL LIBRARY
of Science, Technology, Engineering and Social Studies
The 1000-volume original paperback library in aid of education, industrial training and the enjoyment of leisure
Publisher: Robert Maxwell, M.C.

WEAR OF METALS

INTERNATIONAL SERIES IN

MATERIALS SCIENCE AND TECHNOLOGY

VOLUME 18

Editor: J. HALLING, M.Eng., Ph.D., C.Eng.

OTHER TITLES IN THE INTERNATIONAL SERIES IN MATERIALS SCIENCE AND TECHNOLOGY

KUBASCHEWSKI, EVANS & ALCOCK
Metallurgical Thermochemistry

RAYNOR
The Physical Metallurgy of Magnesium and its Alloys

WEISS
Solid State Physics for Metallurgists

PEARSON
A Handbook of Lattice Spacings and Structures of Metals and Alloys—Volume 2

MOORE
The Friction and Lubrication of Elastomers

WATERHOUSE
Fretting Corrosion

REID
Deformation Geometry for Materials Scientists

BLAKELY
Introduction to the Properties of Crystal Surfaces

GRAY & MULLER
Engineering Calculations in Radiative Heat Transfer

MOORE
Principles and Applications of Tribology

CHRISTIAN
The Theory of Transformations in Metals and Alloys Part 1, 2nd Edition

HULL
Introduction to Dislocations, 2nd Edition

SCULLY
Fundamentals of Corrosion, 2nd Edition

WEAR OF METALS

by

A. D. SARKAR B.Sc., M.Eng., Ph.D.

Principal Lecturer, John Dalton Faculty of Technology, Manchester Polytechnic

PERGAMON PRESS

Oxford · New York · Toronto · Sydney Paris · Frankfurt

U.K.	Pergamon Press Ltd., Headington Hill Hall, Oxford OX3 0BW, England
U.S.A.	Pergamon Press Inc., Maxwell House, Fairview Park, Elmsford, New York 10523, U.S.A.
CANADA	Pergamon of Canada Ltd., P.O. Box 9600, Don Mills M3C 2T9, Ontario, Canada
AUSTRALIA	Pergamon Press (Aust.) Pty. Ltd., 19a Boundary Street, Rushcutters Bay, N.S.W. 2011, Australia
FRANCE	Pergamon Press SARL, 24 rue des Ecoles, 75240 Paris, Cedex 05, France
WEST GERMANY	Pergamon Press GmbH, 6242 Kronberg-Taunus, Pferdstrasse 1, Frankfurt-am-Main, West Germany

First edition 1976

Library of Congress Cataloging in Publication Data

Sarkar, A. D.
Wear of metals.

(International series in materials science and technology; v. 18)
1. Mechanical wear. 2. Friction. 3. Metals.
I. Title.
TA418.4.S27. 1976 621.8′9 75-37574
ISBN 0-08-019738-8
ISBN 0-08-019737-X Flexi

In order to make this volume available as economically and rapidly as possible the author's typescript has been reproduced in its original form. This method unfortunately has its typographical limitations but it is hoped that they in no way distract the reader.

Printed in Great Britain by A. Wheaton & Co., Exeter

TO FRANCES

CONTENTS

PREFACE

Currently, there is a great deal of interest in the wear of metals and materials throughout the world. I have thus ventured to write this book largely aiming at the undergraduates, although I hope that some of the chapters dealing with wear of specific materials will be of use to new research workers. Generally, only the principles have been discussed and specific materials for a particular situation have only been quoted as a means of illustrating a principle. It is assumed that the reader has a background knowledge of metallurgy and engineering principles.

A prerequisite of wear is interaction of surfaces. Surface topography is thus discussed in chapter 2. Wear cannot be isolated from friction and this is discussed in chapter 4, which is preceded by an analysis of contact mechanics in chapter 3. Tribological contacts are both plastic and elastic and the formation of junctions by plastic encounter is high-lighted in chapter 5. Adhesive wear theories are given in chapter 8, preceded by the work of Tomlinson and by running-in wear in chapters 6 and 7. Theories of wear due to oxides and surface contaminants are given in chapters 9 and 10 and chapter 11 discusses abrasive wear. A few chapters are devoted to such aspects as metal transfer, role of crystal structure, rolling resistance and wear. Available literature on fretting has been reviewed together with the friction and wear of some technologically important materials such as cast iron and aluminium-silicon alloys. A chapter on the friction and wear of polymers has been given in a comparative spirit. The final chapter summarises some other types of wear and points out a probable approach towards design against wear. There are excellent text books giving details of materials and manufacturing techniques for tribological components, some examples of which are given in chapter 22.

My thanks are due to the following for permission to reproduce diagrams: Edward Arnold, Oxford University Press; to the editors of Proc Roy Soc., Proc Inst Mech Engrs, Jl Appl Physics., The Engineer, Proc Phys Soc., Cobalt, Wear, Jl Mech Eng Science, Jl Appl Mech and Tribology. I am grateful to Mr. R. M. Jones and to Pergamon for preparing the illustrations. My thanks are also due to Professor J. Halling for reading the manuscript and making valuable suggestions.

It has been difficult to standardise on the mathematical symbols because of the large number involved. The meaning of each symbol, however, has been given as and when it has appeared in the text.

Liverpool A.D.S.

CHAPTER 1

INTRODUCTION

If one of the constituents in the definition of civilisation is the use of technology by homo sapiens, barbarism gave way to the foundation of modern society in a mere 2000 to 3000 years in the neolithic period[1] (ca 3500 to 1500 BC). Technological advancement demands shaping of metals and non-metals to a preconceived design of component, be it a spear for hunting or a wheeled vehicle for transportation. Of the tribological implements, the most common ones were drills made from shell, stone and bones. There is evidence to suggest that stone was a favourite wear resistant material in the middle ages (400-1400 AD) in Europe and has been used in ploughshares or rims of wheels on vehicles. Technologists have always been aware of wear and have sought after fundamental information either to combat the process or to use it to their advantage. Since a prerequisite of wear is that the interacting components must be in intimate contact, engineers have always attempted to separate surfaces by interposing a lubricant between them.

1.1 *Adhesion*

Engineering surfaces are rough and possess hills and valleys so that contact between two solids occurs only at a few isolated points resulting in a true area of contact which is a fraction of the apparent area. The applied normal stress is therefore very high in the regions of contact and may exceed the yield point of one or both of the solids. The contact areas will then weld together forming junctions. These must be broken to initiate and sustain relative motion and the force necessary to disrupt the junctions is a measure of friction.

1.1.1 *Contact Resistance.* Naturally, attempts have been made to measure the true area of contact experimentally, one method being the determination of electrical resistance between two contacting solids. Suppose that two solids A and B each of specific conductivity λ make a circular contact of diameter 2a. Let the surface contaminant on the solid have uniform thickness and let it have a resistance per cm^2 of σ. The combined resistance R of the junction is then given approximately as

$$R = \frac{1}{2a\lambda} + \frac{2\sigma}{\pi a^2} \qquad (1.1)$$

To measure the contact resistance, it is necessary to use the current-potential method as R is small compared with the resistance of the leads. If conditions are clean, the second term in equation (1.1) can be neglected and the value of a can be obtained from equation (1.1). The true area of contact can also be obtained by microscopic observation or indirectly from a knowledge of the flow stress σ_y of the metal, since for an applied normal load W, $\pi a^2 = W/\sigma_y$.

1.2 *Contaminants*. As the junctions break, wear debris is expected. To prevent wear, therefore, all that should be necessary is to forestall the formation of junctions. This can be done by partitioning the interface with an intruding layer which impedes the interaction of metal atoms at the contact spots of two metallic solids. It is fortuitous that surfaces exposed to room atmosphere are quickly covered with a layer or more of sorbed gases and oxides. Oxidation should be facilitated by frictional heating and contaminants minimise the wear of many apparently unlubricated systems. The sorbed gases may vaporise if overheated and the oxide layers may disrupt if the normal stresses are high, exposing nascent metal which should increase wear propensity.

To understand about wear, it is imperative to study the topography and the physico-chemical nature of surfaces. The degree of surface and subsurface deformation of solids play a fundamental role in the mechanisms of friction and wear so that the contact stresses and the types of motion must be studied when analysing the nature and amount of wear of the kinematic chains of machinery.

1.3 *Types of Wear*

A potential wear situation exists whenever there is relative motion between two solids under load. Broadly speaking, the motion can be unidirectional or reciprocating either sliding or rolling. There may be a combination of rolling and sliding or wear may occur due to oscillatory movement at small amplitudes. A metal can interact with a non-metal or liquids such as lubricating oil or marine water. Depending on the nature of movement or of the media involved in an interaction under load, the following types of wear have been classified.

1.3.1 *Adhesive Wear*. In this, the relative movement can be unidirectional or reciprocating sliding or interaction occurs under small amplitude oscillatory contact under load. The mating surface peaks are known to flow plastically and to form strong work hardened junctions. As these break under an imposed tangential traction, material loss from the solids may occur.

1.3.2 *Abrasive Wear*. Abrasive particles in the form of wear debris or adventitious particles of grit and dust from the surroundings remain trapped at the sliding interface and remove material largely by ploughing.

1.3.3 *Other Forms of Wear*. Fretting is a form of wear which occurs as a result of oscillatory movement between two surfaces as in machine parts undergoing vibration. Fatigue wear arises as a result of cyclic loading, for example, in rolling element bearings and loss of material occurs by spalling of surface layers. Erosive wear results when grits impinge on solids while cavitation erosion may arise when a component rotates in a fluid medium.

1.4 *Friction and Wear Experiments*

Although information on tests carried out on actual engines for wear data are available, laboratory studies are carried out under controlled conditions, simulating motions encountered in actual situations. Friction and wear results are often obtained from the same machine.

There are a number of machines used to evaluate friction and wear and the principles of a few of them are outlined here. The first requirement is to design a couple which will provide a rubbing interface. This can be done by attaching a pin on a load bar, the pin being in the form of a rod or a hemisphere. The load bar is designed to carry the normal load but its deflection due to frictional drag between the pin and the opposing surface can be measured and suitable calibration provides the value of friction. Some typical principles are shown in Figs. 1.1 - 1.4 where the pin runs on a reciprocating table (Fig. 1.1) or on a bush (Fig. 1.2). Figure 1.3 shows the pin to be in the form of a cylinder placed across another larger cylinder. Figure 1.4 shows a cylindrical or hemispherical pin rotating on a bush or a cylinder. The deflection of the load bar can be measured by using transducers or strain gauges.

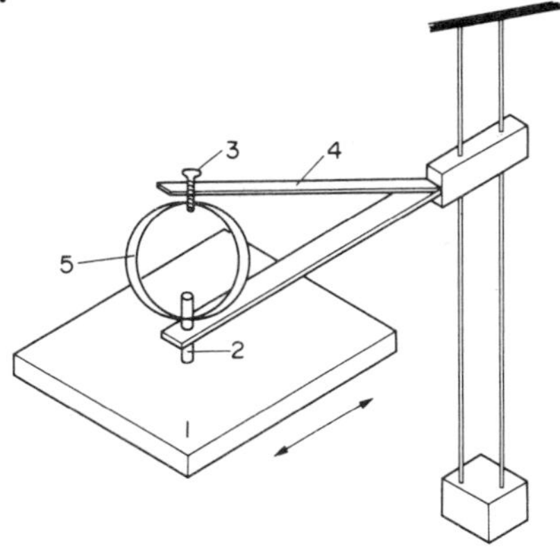

Fig. 1.1 An apparatus to measure friction at light loads (schematic). (1) A flat surface capable of motion in the directions of the arrows. The pin (2) is attached to a ring (5). Normal load applied through the ring by a screw (3) and the load is evaluated by the deflection of the ring. A horizontal arm (4), attached to a bifilar suspension is attached to the ring (5) and the frictional drag causes it to deflect which is measured.

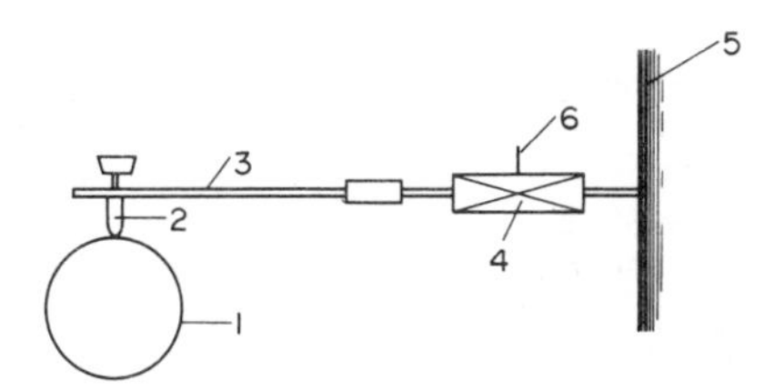

Fig.1.2 A pin bush machine. The pin (2), attached to a horizontal load bar (3) runs on a rotating bush (1). The load bar is attached to a spring (4) whose other end is fixed to a rigid support (5). Frictional drag causes an extension of the spring. A needle (6) fixed to the spring can be attached to a soft iron suspended in a magnetic field. Deflection of the needle due to friction will cause the iron core to move resulting in a change in the induced field strength. The resulting signal can be amplified and recorded.

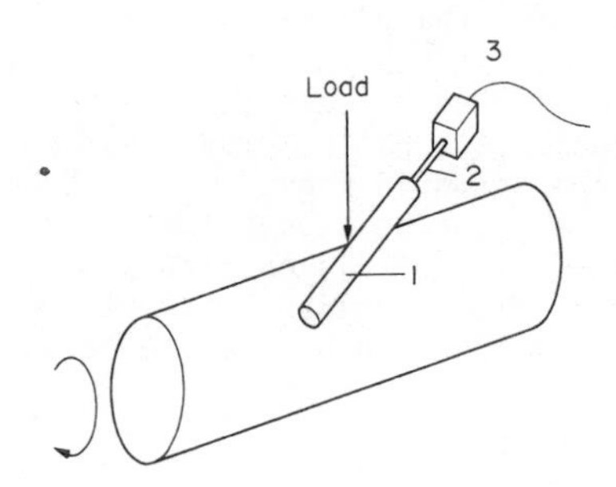

Fig. 1.3 A crossed-cylinder (1) mounted on a spring (2). Movement due to frictional drag measured by a transducer (3).

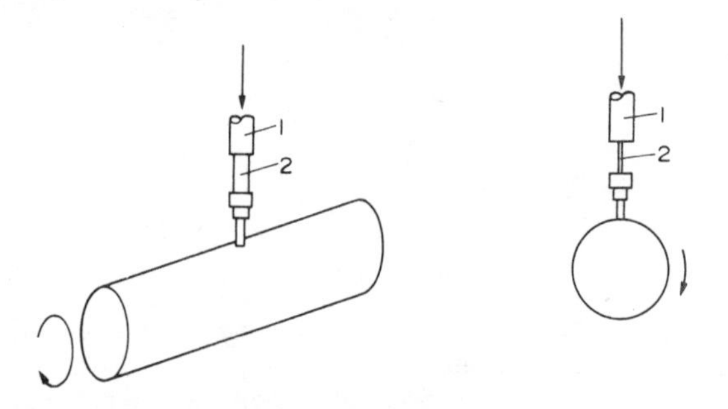

Fig. 1.4 A load bar (1) with strain gauges mounted on the reduced section (2). Bending of the bar due to friction unbalances a bridge circuit.

Friction and wear are also measured in a four-ball machine or by disc tests. The primary purpose of the former is to evaluate the anti-seizure properties of oils but it can also be used as a wear tester. Three 12.5 mm diameter contacting steel balls are held in a ring and a fourth ball is fixed at the end of a vertical shaft and made to contact the lower three balls. A lever carrying an adjustable load presses the three stationary balls against the fourth and the torque transmitted is measured during rotation of the vertical shaft. The coefficient of friction can be recorded throughout a test and the wear scars on the three balls are measured.

In the disc machine a disc under load is made to rotate against another and is normally used to simulate studies of gear design. One disc is carried in a rigid bearing while the other is supported by a swinging device.

The pin-disc machine is a popular wear testing apparatus (Fig. 1.5) where the pin is loaded normally. The variables are normal load, sliding velocity, atmosphere and the temperature of the environment. The amount of wear can be established by weighing the pin with a micro-balance. A complete wear test involves plotting weight loss against sliding intervals to obtain a running-in and a steady state wear as shown in Fig. 7.1, chapter 7. It is important to work under chemically clean conditions and the method is tedious. An alternative method is to use a conical pin and mount it on a hinged load bar. The pin is run and, after stopping the machine, the load bar is swung over by 180° (Fig. 1.6) and the diameter of the wear scar is noted so that, knowing the angle of the cone, the volume loss can be calculated.

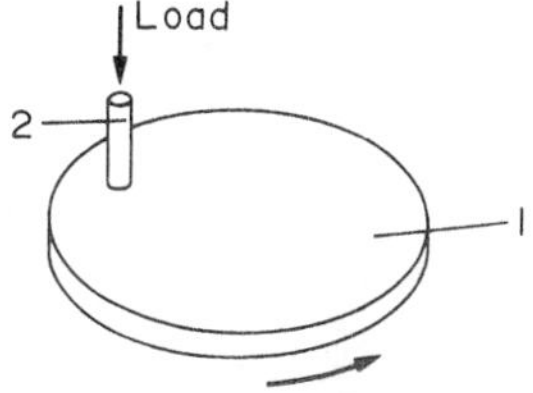

Fig. 1.5 A pin-disc machine (1) Disc; (2) Pin

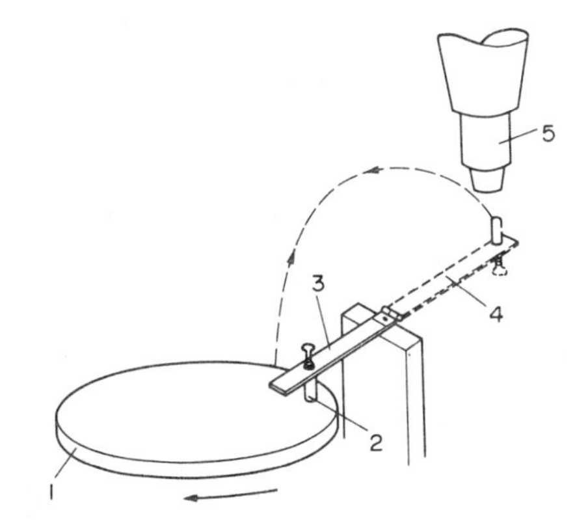

Fig. 1.6 A hinged load bar (1) Disc; (2) Pin; (3) Load bar; (4) Load bar swung through 180° to measure wear scar by the microscope (5).

The crossed cylinder configuration shown in Fig. 1.3 is also used for wear experiments. The wear scar on the cylindrical pin is elliptical and the minor and major axes can be measured with a microscope. Halling[2] has shown that if the depth of the wear scar is small and if 2r is the minor diameter of the generated ellipse, the volume V of metal removed is given by

$$V = \frac{\pi}{4} \left(\frac{r_s}{r_b}\right)^{\frac{1}{2}} r \qquad (1.2)$$

where r_s and r_b are the radii of the rotating and the stationary cylinder respectively.

The wear machines can be enclosed in a chamber for experiments in a controlled atmosphere but, if the couple are to be completely free from contaminants, it is necessary to employ a vacuum of the order of 10^{-12} torr. This is expensive but Halling's idea[2] of using a lathe (see chapter 16) is a very ingenious and cheap way of generating clean surfaces by machining the rotating cylinder with a cutting tool.

1.5 *Metallurgical Examination*

The surface texture is an important parameter and chapter 2 discusses surface topography and the use of a profilometer to plot the hills and valleys. The prior surface of a metal, however, continues to change as a result of deformation a d wear and this can be usefully followed by optical and electron microscope or with the aid of a stereoscan. It is important to observe the nature of the subsurface and apart from microscopic observation, microhardness or electron diffraction technique provides valuable information. Dissimilar metals in rubbing contact may result in the formation of intermetallic compounds giving a composition change at and below the surface. A microprobe analysis on tapered sections of pins is a valuable aid to establish the chemical composition which may also be affected as a result of oxidation. The wear debris is often subjected to X-ray diffraction analyses to obtain information regarding metallurgical interaction between the members of a couple.

1.6 *Application of Wear Results*

There is often apparent disagreement between theoretical and experimental studies of wear rates of metals and materials and the application of these results to industrial situations. This need not be so and what is desirable is a spirit of participation among the two.

Whenever a new material is investigated, the first phase of the work is usually an evaluation of the amount of wear with increasing sliding distance. Taking the slopes of the steady wear, the next step is to plot wear rates against load. At low loads, there is a regime of mild wear, followed by a transition regime when the rate of wear may increase by a few orders of magnitude (see Fig. 23.3 or 25.1). Workers find it difficult to resist the temptation to keep on increasing the load until a complete wear rate load curve has been established. However, one does not expect a machine to operate under conditions beyond the regime of mild wear. This is not to say that the knowledge of the transition load is unimportant but future effort should concentrate on the mechanism of mild and running-in wear to provide information for design engineers.

There is an understandable scepticism in certain sections about using laboratory data for design work obtained from simulated studies using the types of machines outlined in section 1.4. An attempt to overcome this is perhaps the concept of the IBM wear model[3] which attempts to obtain empirical expressions for specific geometry which could be applied to design work. Firstly a single pass a is defined which is the apparent length of contact. Thus for a pin-bush machine with pin and bush diameters as a and r respectively,

$$\text{a single pass} = a$$

Number of passes for one complete revolution is $2\pi r/a$.
The amount of wear which is less than the surface finish of the component under investigation is called zero wear.

For zero wear, in a single pass,

$$\tau_{max} < 0.54\ \tau_r$$

where τ_{max} = maximum shear strength experienced

τ_r = shear strength necessary to cause yielding

Measurable wear, i.e. wear greater than the surface finish is given as

$$d\left[\frac{q}{\left\{\tau_{max}{}^{r}\right\}\frac{9}{2}}\right] = CdM$$

where M is the total number of passes, C is a constant and q is the amount of wear which is obtained by feeding in the values of the constants and integrating the above expression.

The IBM model is perhaps valuable to provide data for individual situations but laboratory studies using pin-disc and other configurations must remain important because of the scope they offer to obtain basic understanding of the wear processes in metals.

Another objection is that laboratory wear studies are accelerated tests but so are creep and fatigue testing of metals and they have provided much wanted information to design engineers.

The problem is that it is seldom possible to state with confidence that only a single mode of wear is responsible in a particular plant. For example, to avoid unwanted abrasive wear, apart from efficient external seals one must consider the geometry of the interface to assess if the debris can escape from the interface as it is generated. The compatibility of the lubricant and the materials of the components from the viewpoint of corrosion needs taking into account. Clearly, research and application of the results are not separated by a chasm. No doubt, there are certain grey areas between them but the aim should be to make any boundary diffuse by an interchange of experience between research workers, manufacturers and users of tribological components.

REFERENCES

1. Dowson D, *'Tribology before Columbus'*, Mechanical Engineering, (1973).
2. Halling J, *Wear*, (1961), *4*, 22.
3. Bayer R G, Clinton W C, Nelson C W and Schumacher R A, *Wear*, (1962), *5*, 378.

(See also Halling J (Ed), *'Principles of Tribology'*, The Macmillan Press (1975).

CHAPTER 2

SURFACE TOPOGRAPHY

A knowledge of the prior nature of surfaces is very important to understanding the mode of interfacial interaction between moving parts of machinery. An important aspect of surfaces is whether they are free from atmospheric contaminants or oxides. On the other hand, it may be necessary to know whether a surface is mechanically soft. Above all, the lack of flatness of a surface on a microscopic scale has been shown to be the basic premise with regard to the understanding of the mechanism of friction and wear.

2.1 *Asperities*

A perfectly flat surface as shown in Fig. 2.1a is never achieved by the usual methods of surface preparation such as machining, grinding, lapping etc. Instead, surfaces have asperities, that is undulations in the form of hills and valleys. Fig. 2.1 shows that the surfaces b - e have protuberances which are all of the same depth but the wave length is macro-geometrical as in b or micro-geometrical (fig. 2.1e). Figure 2.1b shows that the asperities have a large wave length and such a surface is described as smooth but not flat. Figure 2.1e has many protuberances of short wave length and a surface like this is deemed to be flat but rough. The nature of surfaces with roughness in between Fig. 2.1b and e is shown in Fig. 2.1c and d. In actual engineering surfaces, depending on the method of production, the height of the peaks may vary between 0.05 μm to 50 μm while the spacings between them range from 0.5 μm to 5 mm.

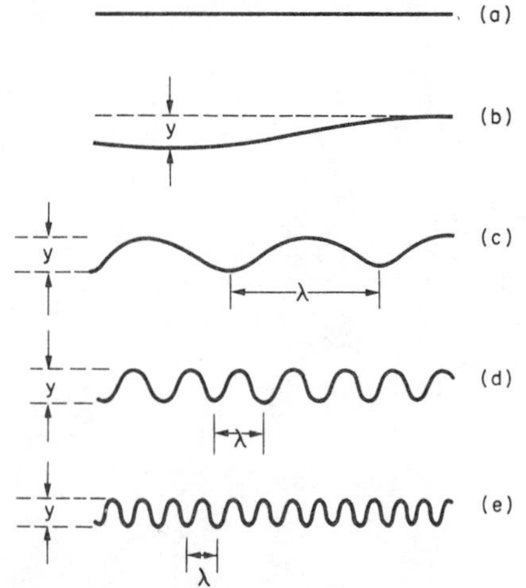

Fig. 2.1 Idealised traces of surface asperities showing the same peak to valley height but differing wavelengths.

2.2 *Measurement of Waviness*

Because of the rapidity with which a record of surface micro-texture can be obtained, the Taylor Hobson Talysurf is widely used. The principle of the instrument is to move a stylus over a representative length of the surface under examination (Fig. 2.2). Figure 2.2a shows a stylus moving over a rough surface with the result that the oscillatory movement causes the armature to tilt. The armature supports the pole pieces around which are two coils I_1 and I_2, each carrying a high frequency current. The air gap between the armature and the pole pieces varies as the stylus traverses the surface under examination. This results in a variation of the impedance of the coils and hence the magnitude of the high frequency current. The coils I_1 and I_2 are shown in the schematic electrical circuit as part of an a.c. bridge. The bridge is supplied with a high frequency current by a valve oscillator O and the modulated current is amplified by an amplifier A. A demodulator B demodulates the output so that the fluctuating current as a result of the unbalance of the bridge circuit actuates the pen recorder with the frequency of the carrier current from the oscillator eliminated.

The unbalance of the bridge results due to the tilting of the armature. That is, the mechanical movement is converted to an electrical quantity and the surface undulation is traced by the pen recorder.

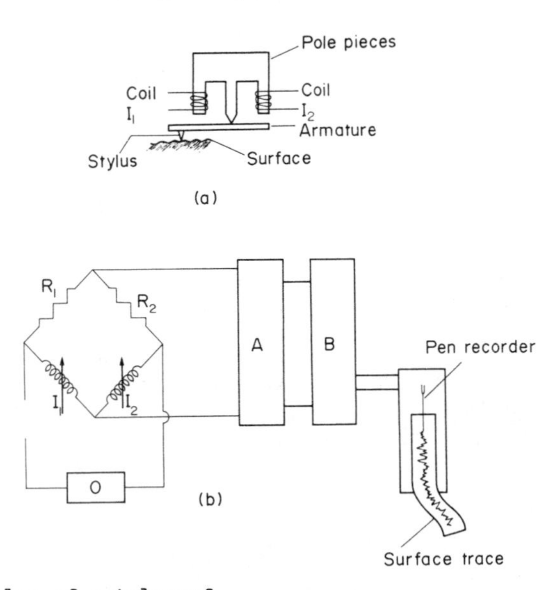

Fig. 2.2 Principles of a talysurf.

Since the peaks and the spacings between them are small, the traces are magnified in the range 400 to 75000 times. The length traversed by the stylus is very large compared with the dimensions of the peaks and valleys. To avoid the width of the recording paper to be unmanageably wide, therefore, the horizontal magnification is much less than that in the vertical direction. A typical trace is shown in Fig. 2.3a where the ratio of the horizontal to the vertical scale is 1:100. The problem, of course, is that the shapes of the hills and valleys are misrepresented and the peaks appear more steep than they really are. Figure 2.3.b shows the true nature of the surface waviness where a small segment marked x in Fig. 2.3.a is re-drawn with both axes at the same magnification.

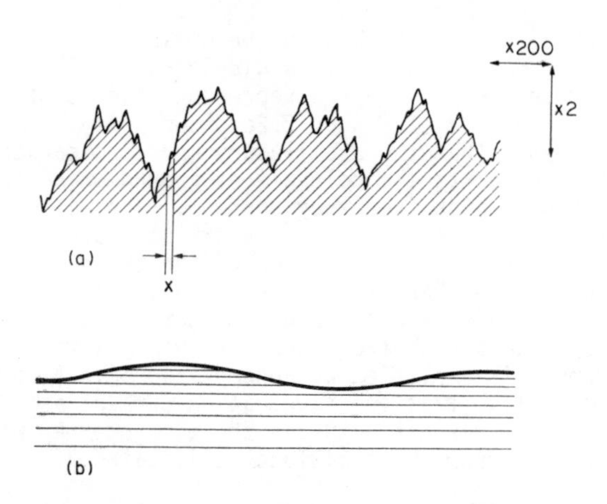

Fig. 2.3 A profilometric trace of a surface (not an actual trace). (a) As obtained with a talysurf where the peaks and spacings between them are distorted, since the horizontal magnification is 100 times less than the vertical. (b) Showing how a small width x, when replotted with the same horizontal and vertical magnifications, is really smooth.

2.3 *Asperity Angle*

Halliday[2] has measured the angles of surface asperities with the aid of reflection electron microscopy by viewing the surface as obliquely as possible so that the asperities are seen in profile. Abraded surfaces often show an asperity to slope at different angles to the surface from two directions. Figure 2.4 shows the nature of the asperity slopes where the angle α falling ahead of the abrading particle is always greater than β. For copper, α and β were found to be 2.5 and 1° respectively, being much smaller than this, about 0.1°, for tool steel. Abraded mild steel showed α and β to be the same at 2°. Asperity angles play a positive role in the mode of deformation of metals when relative movement of surfaces under load ensues and often an average value of 5° is taken, assuming $\alpha = \beta$.

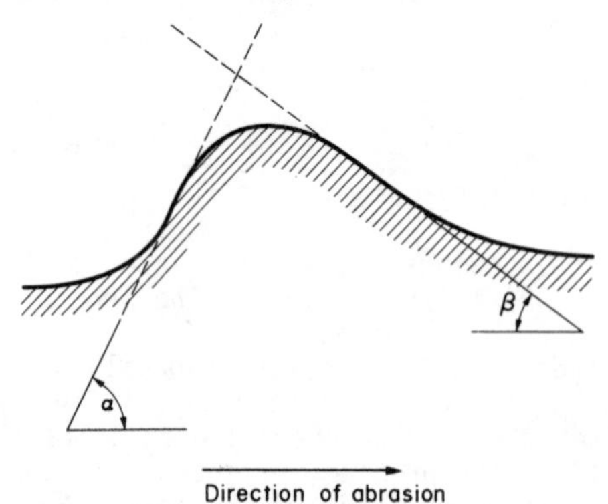

Fig. 2.4 Asperity slopes of an abraded surface. (after Halliday[2]).

2.4 *Measure of Roughness*

Roughness of surfaces has been described in the following way:
(1) The maximum peak to valley height, R_t (Fig. 2.5)
(2) The centre line average, cla or R_a and the root mean square, R_s.
(3) The bearing curve.

As shown in Fig. 2.1, surfaces possessing various degrees of roughness can have the same value of R_t, so that measurement of the peak to valley height does not completely describe a surface. It is important to have an estimate of the wavelength and the spacings between asperities have been specified by stating the number of peaks for a given length. An average peak to valley distance can also be quoted and this is referred to as the Ten Point Height, R_z, which is the arithmetic mean of the heights of the five highest peaks and the same number of deepest valleys. A common method of expressing the average height of surface undulations is to state R_a or R_s. The root mean square is popular in the USA while there is a preference for cla value in the UK.

Taking y as the height coordinate and x as the horizontal coordinate (Fig. 2.5), for a length ℓ,

$$R_a = \frac{1}{\ell} \int_0^{\ell} /Y/dx \tag{2.1}$$

That is, referring to Fig. 2.5, the average height of the peaks is obtained by adding the shaded areas and then dividing it by ℓ. The areas can easily be measured by a planimeter.

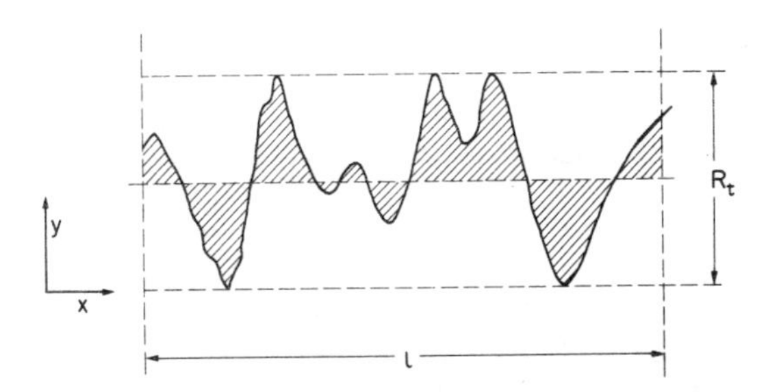

Fig. 2.5 Surface asperities. R_t = maximum peak to valley height.

The root mean square value is given by the following relationship:

$$R_s = \left(\frac{1}{\ell} \int_0^{\ell} y^2 dx \right)^{\frac{1}{2}} \tag{2.2}$$

2.5 *Fullness or Emptiness*

Figure 2.6 shows three asperities of equal heights enclosed within a rectangle ABCD. If an object is now placed on AB, it is seen that the load is supported only by the peaks of the three asperities. An estimate of the fractional area carrying the load is given by a form factor K and in Fig. 2.6

$$K = \frac{\text{Area of shaded portion}}{\text{Total area of rectangle}} \tag{2.3}$$

K is referred to as the degree of fullness and the degree of emptiness

$$K_p = 1 - K \tag{2.4}$$

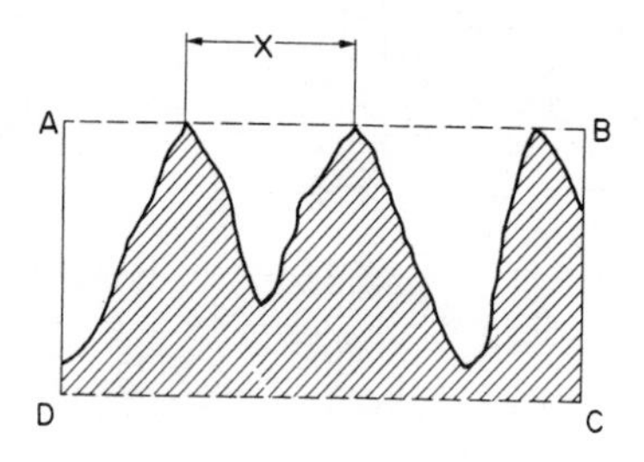

Fig. 2.6 An idealised trace to show fullness and emptiness.

2.6 *Abbot's Bearing Curve*

Suppose a surface under load on the asperity tips is slid in the horizontal direction resulting in wear of part of the shaded area. As in a bearing, if sliding is continued, progressive wear will occur eventually removing all the asperities. An idealised bearing curve is shown in Fig. 2.7. Suppose the tips have worn to a depth x leaving flats of width a_1 and c_1. Add a_1 and c_1 as shown in the right hand diagram at a depth x from the top. For the depth y, add a_2, b_2 and c_2 and continue this way until the bearing curve is completed. This type of information is of interest when surfaces are examined after they have been subjected to relative movement under load.

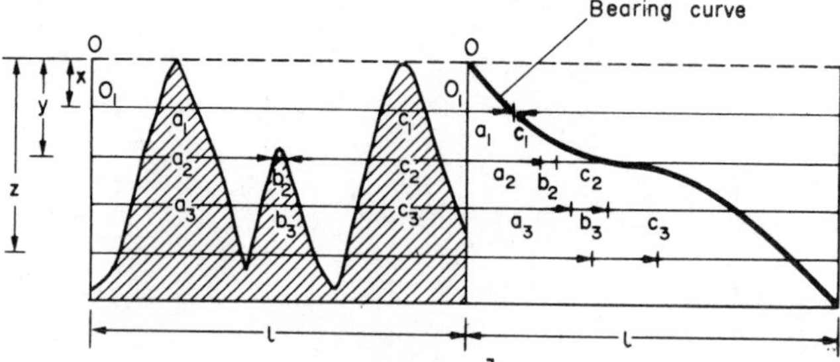

Fig. 2.7 Bearing area curve (after Miller[1]).

Most engineering surfaces tend to give a Gaussian distribution of texture heights (Fig. 2.8a). Abbot's bearing area curve, in effect, is the cumulative distribution (Fig. 2.8b) of the all ordinate distribution curve which is

$$\Psi(z) = \int_{-\infty}^{+\infty} \phi(z)dz \tag{2.5}$$

where z is the profile height measured from a reference plane and $\phi(z)$ is the probability density function of the distribution of these heights.

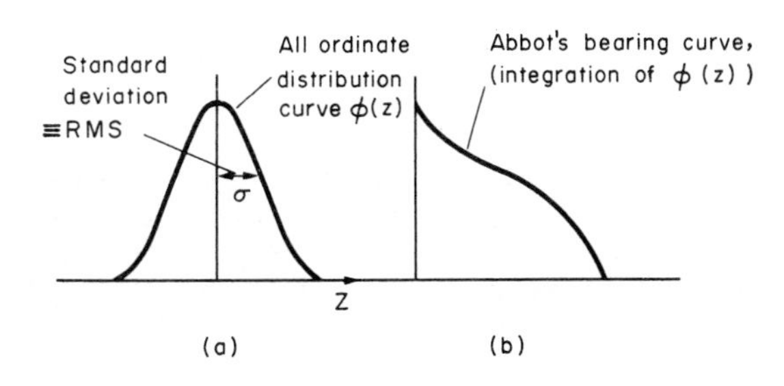

Fig. 2.8 Shows Abbot's bearing area curve (b) to be the cumulative distribution of the all ordinate distribution curve (a).

REFERENCES

1. Miller L, *'Engineering Dimensional Metrology'*, Edward Arnold Ltd. (1962).
2. Halliday J S, *Proc Inst Mech Engrs*, (1955) *169*, 777.

CHAPTER 3

CONTACT OF SOLIDS

Chapter 2 showed that if a surface is pressed upon another, the load is supported on the tips of a few peaks of the bottom surface, assuming that the top member of the couple is perfectly flat. There is thus an apparent area of the interface between two surfaces but the true area of contact is only at a few points at the tips of the asperities. If the load is low and the material has a high yield stress, contact will be elastic. On the other hand, for the opposite case, the interface will flow plastically. Both modes of behaviour appear to occur in the kinematic interface of most tribological situations and both friction and wear processes in metals and materials have a dependence on the nature of the true area of contact of the mating surfaces.

3.1 *Single Contact*

Consider the case (Fig. 3.1) of a hard hemispherical slider loaded against a soft flat surface. It is assumed that both the rider and the flat surface are perfectly smooth, that is, they do not possess undulations. Depending on the load, the rider will indent elastically a circular area of diameter 2a. Hertz (1896) in his classical analysis has shown that, for elastic contact, the compressive stress σ_r at any radial distance r from the centre of the indented area is given by

$$\sigma_r = \sigma_{max}\left(1 - \frac{r^2}{a^2}\right)^{\frac{1}{2}} \qquad (3.1)$$

This means that provided the yield stress of the soft material is not exceeded, the maximum compressive stress is at the centre of the circle of contact, falling to zero at the edge where r = a (equation 3.1). The variation of the compressive stress across the diameter of indentation is shown schematically in Fig. 3.1

If W is the applied load

$$\sigma_{max} = \frac{3W}{2\pi a^2} \qquad (3.2)$$

Hertzian analyses also show that, although the maximum normal stress is on the surface, the peak shear stress τ_{max} is within the material at a distance 0.5a below the surface (0, Fig. 3.1) and

$$\tau_{max} = 0.31\sigma_{max} \qquad (3.3)$$

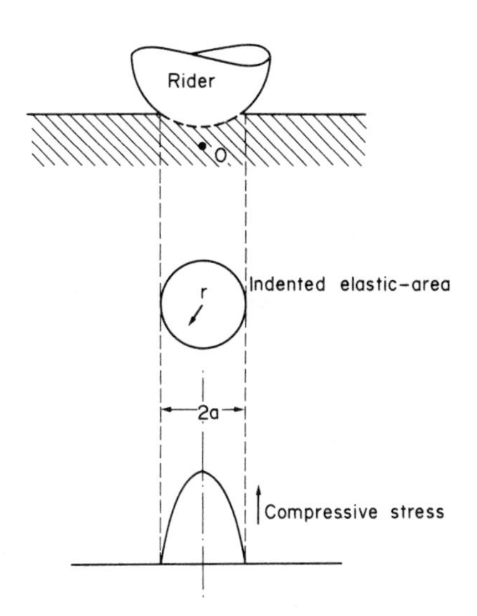

Fig. 3.1 A hemispherical rider pressing on a flat surface.

3.1.1 *General Case*. Generally, for two curved surfaces of radii of curvature R_1 and R_2 with Young's moduli E_1 and E_2 and Poisson's ratios υ_1 and υ_2 respectively, the elastic contact radius a_g is given by

$$a_g = \left(\frac{3WR}{4E} \right)^{\frac{1}{3}} \tag{3.4}$$

The maximum Hertzian pressure, σ_{mg} is

$$\sigma_{mg} \simeq \frac{2}{3} \left(\frac{WE^2}{R^2} \right)^{\frac{1}{3}} \tag{3.5}$$

where $$\frac{1}{E} = \frac{1 - \nu_1^2}{E_1} + \frac{1 - \nu_2^2}{E_2}$$

and $$\frac{1}{R} = \left| \frac{1}{R_1} \right| + \left| \frac{1}{R_2} \right|$$

The modulus sign signifies that for counterformal contact such as a sphere resting on a sphere, both R_1 and R_2 are positive. If contact is conformal such as a ball and a cup, the radius of curvature of the cup is negative.

The area of elastic contact A_e is

$$A_e = \pi a_g^2$$

or from equation 3.4,

$$A_e = \pi\left(\frac{3WR}{4E}\right)^{\frac{2}{3}} \quad (3.6)$$

The mean pressure $\bar{\sigma}$ increases over this area up to a limiting load at which the elastic limit of the soft material is exceeded. Yielding of the material should initiate at a point 0.5a below the centre of the indentation where the shear stress is maximum.

3.2 *Multiple Contact*

Engineering surfaces, however, as presented in the case of a hemispherical rider pressing on a soft surface, are never perfectly smooth. In fact, the sphere will have asperities and these in turn will be covered with micro-asperities. Rather than a smooth contact area of radius a as in Fig. 3.1, the sphere will rest on a few asperities on the flat surface. Whereas the elastic contact area is shown to be proportional to the cube root of the square of the load (equation 3.6), Archard[1] shows that if the protuberances on the asperities are considered and the roughness of these protuberances themselves is taken into account and this breakdown is carried through successively (Fig. 3.2), a stage is reached when the area of elastic contact becomes very nearly proportional to the applied normal load.

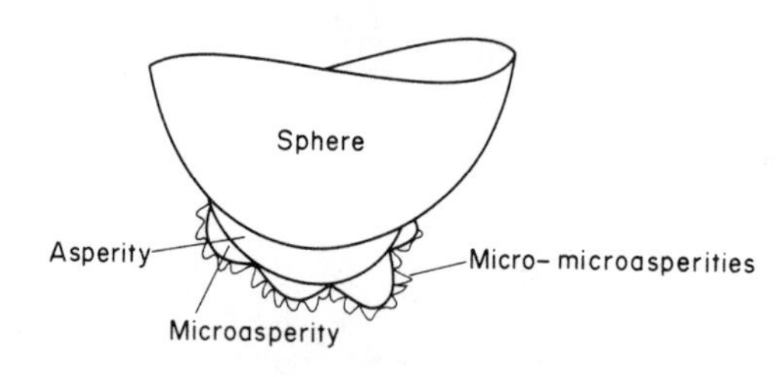

Fig. 3.2 A sphere with micro-microasperities.

3.3 *An Idealised Rough Surface*

Ignoring the presence of microasperities consider[2] a rough surface with perfectly spherical asperities each of radius R (Fig. 3.3). Let a perfectly smooth surface penetrate these protuberances and let the applied normal load be such that deformation is wholly elastic. The height of each asperity from an arbitrary reference line XX is z and the compliance δ equals (z-d). Assume that there is no interaction between the protuberances so that the deformation of one asperity does not affect the heights of the others. By assumption, all the asperities deform equally, that is, they move down simultaneously through a distance (z-d) under a normal load W. If there are n asperities per unit area, the uniformly distributed total load supported by the rough surface is nW_i, where W_i is the load sustained by each asperity. The load-area relationship has already been stated in 3.1.1. In terms of the compliance δ, from Hertzian analysis, the elastic contact radius a of the indentation by each asperity is,

$$a = R^{\frac{1}{2}}\delta^{\frac{1}{2}}$$

The area of contact A_{ei} is

$$A_{ei} = \pi R\delta$$

The load W_{ei} is

$$W_{ei} = \frac{4}{3} ER^{\frac{1}{2}}\delta^{\frac{3}{2}}$$

or, putting $\delta = (z-d)$

$$W_{ei} = \frac{4}{3} ER^{\frac{1}{2}}(z-d)^{\frac{3}{2}} \tag{3.7}$$

and

$$A_{ei} = \pi R(z-d) \tag{3.8}$$

The total load W_e and the total area of true contact A_e are given by

$$W_e = \frac{4}{3} E\ n\ R^{\frac{1}{2}}\left(\frac{A_{ei}}{\pi R}\right)^{\frac{3}{2}}$$

Since the true contact area, $A_e = nA_{ei}$,

$$W_e = \frac{\frac{4E}{3}}{3\pi^{\frac{3}{2}}n^{\frac{1}{2}}R} A_e^{\frac{3}{2}}$$

That is,

$$A_e \propto W_e^{\frac{2}{3}} \tag{3.9}$$

Equation 3.9 shows that, when contact is elastic, the true area of contact is not directly proportional to the applied normal load.

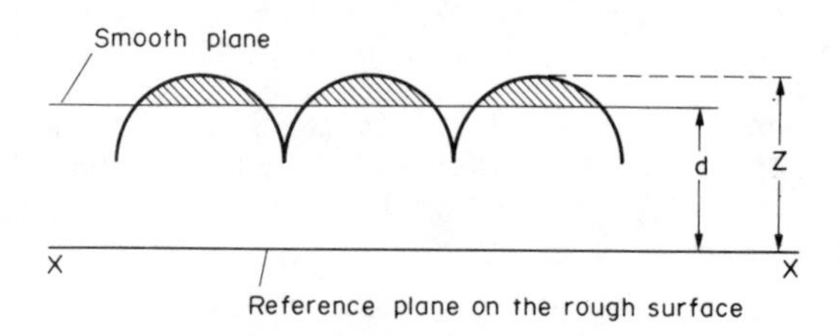

Fig. 3.3 An idealised rough surface being penetrated by a flat surface. (after Halling[2])

3.4 *A Realistic Rough Surface*

The previous analysis is based on an idealised asperity distribution but true surfaces have asperities varying in height and are distributed randomly shown schematically in Fig. 3.4. Let the lower surface be penetrated by another which is nominally flat. The summits of the asperities which have been penetrated, shown shaded in Fig. 3.4, are considered spherical, each having the same radius. However, since the heights vary randomly, a probability statement is made to estimate the number of asperities on the surface which are in contact with the opposing member of the couple.

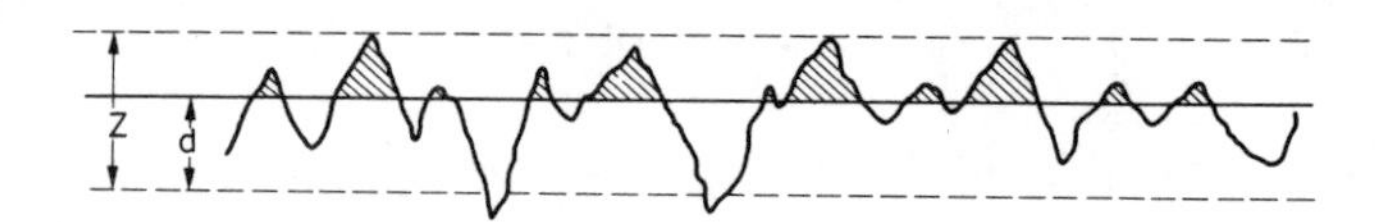

Fig. 3.4 A realistic surface being penetrated by a flat surface (after Greenwood and Williamson[3])

As the two solids approach under a normal load, at equilibrium, the compliance is $\delta = (z-d)$, that is, those asperities whose heights are greater than d will be penetrated. The probability of contact at an asperity with height z is stated as

$$\text{prob}\ (z>d) = \int_d^\infty \phi(z)dz \qquad (3.10)$$

where $\phi(z)$ is the probability distribution curve.

If there are n asperities per unit area, the expected number of contacts, N, is given by

$$N = n \int_d^\infty \phi(z)dz \qquad (3.11)$$

or from equations 3.8 and 3.11, the expected total area of contact is

$$A_e = n\pi R \int_d^{\infty} (z-d)\phi(z)dz \tag{3.12}$$

If the asperities are assumed to be spherical with the same radius, each indented area should be the same for a particular compliance. It is, perhaps, more realistic to assume a mean asperity radius or a mean contact area.

Similarly, the load W_e is given by (from equation 3.7 and 3.11)

$$W_e = \frac{4}{3} nER^{\frac{1}{2}} \int_d^{\infty} (z-d)^{\frac{3}{2}} \phi(z)dz \tag{3.13}$$

It is usual to introduce standardised variables. If A is the nominal contact area and σ the standard deviation of the height distribution, putting $h = d/\sigma$ and $s = z/\sigma$

$$N = nAF_0(h)$$

$$A_e = \pi nAR\sigma F_1(h)$$

$$W_e = \frac{4}{3} nAER^{\frac{1}{2}}\sigma^{\frac{3}{2}} F_{\frac{3}{2}}(h)$$

where

$$F_m(h) = \int_h^{\infty} (s-h)^m \phi^*(s)ds$$

where $\phi^*(s)$ is the standardised height distribution scaled to make its standard deviation unity.

3.4.1 *Exponential Distribution.* The height distribution of most engineering surfaces is Gaussian but an exponential distribution is a fair approximation for the uppermost 25% of most surfaces. Considering the summits only, which make contact when a load is applied,

$$\phi^*(s) = e^{-s}$$

$$F_m(h) = m!e^{-h}$$

This gives

$$N = nAe^{-h}$$

$$A_e = \pi(nR\sigma)Ae^{-h}$$

$$W_e = \pi^{\frac{1}{2}}(nR\sigma)E(\sigma/R)^{\frac{1}{2}}Ae^{-h}$$

comparing A_e in respect of W_e, it is seen that

$$A_e = \text{const} \times W_e \tag{3.14}$$

That is, although the nature of contact is elastic, the true contact area is directly proportional to the applied normal load.

3.5 *Plastic Contact*

Chapter 4 will show that the laws of friction assumes plasticity of the area of contact in a sliding situation. The truth, probably, is that, during the initial traversals of two components under load, the predominant mode of deformation is plastic but the component work hardens to a finite depth below the surface, giving rise to a largely elastic situation. Such phenomena have led to the concept of plasticity index and this is discussed in chapter 14.

For plastic deformation of asperities, the true area of contact A_{pi} and the normal load W_{pi} are expressed[2] in terms of the compliance as

$$A_{pi} = 2\pi R(z-d)$$

$$W_{pi} = \pi HR(z-d)$$

where H is the contact hardness of the material.

Using equation 3.11, it follows that the total area of true contact is,

$$A_p = 2\pi nR \int_d^{\infty} (z-d)\phi(z)dz$$

and the total load, $W_p = 2\pi nRH \int_d^{\infty} (z-d)\phi(z)dz$

Comparing A_p and W_p,

$$\frac{W_p}{A_p} = H \tag{3.15}$$

It should be noted that in later sections H is replaced by σ_y, the flow pressure of the material. An example is equation 4.2 where, strictly speaking, a constant should be incorporated, since $H = \text{constant} \times \sigma_y$.

Equation 3.15 shows that the true area of plastic contact is directly dependent on the load and is independent of the peak height distribution.

3.6 *Effect of Work Hardening*

An elastic confrontation is expected of solids loaded below their yield strength. If, however, the applied normal stress causes plastic flow of one of the surfaces, work hardening will follow with most metals. In that situation, future encounters between the two solids are elastic in nature. The effect of work hardening on the true area of contact has been analysed by Halling and Nuri[4], again assuming surface asperity distribution as in Fig. 3.3. Thus, for a material subjected to a compressive load, the true stress $\bar{\sigma}$ and the corresponding strain $\bar{\varepsilon}$ can be related as

$$\sigma = B\,(\bar{\varepsilon})^{p} \qquad (3.16)$$

where B is a constant and the exponent p is a measure of the degree of work hardening. At p = 1, the situation is elastic while, when p = o, the behaviour is perfectly plastic and B in that case is the yield stress of the metal.

If W is the applied normal load,

$$\bar{\sigma} = \frac{W}{C\pi a^2} \qquad (3.17)$$

where $\frac{1}{C}$ is the constant of proportionality and 2a is the diameter of the circular indentation. C has a value of about 2.8 and it is further assumed that

$$\bar{\varepsilon} = D\,\frac{a}{R} \qquad (3.18)$$

where R is the radius of an asperity.

Combining equations 3.16, 3.17 and 3.18,

$$WR^{p} = Ka^{(p\,+\,2)} \qquad (3.19)$$

where

$$K = \pi BCD^{p} \qquad (3.20)$$

Since, B C, p and D are constants, K is invariant for a given material but only for particular values of B and p.

From the relationship 3.20,

$$D = \left(\frac{K}{\pi BC}\right)^{\frac{1}{p}}$$

Substituting this in equation 3.18,

$$\bar{\varepsilon} = \left(\frac{K}{\pi BC}\right)^{\frac{1}{p}} \frac{a}{R} \qquad (3.21)$$

As in equation 3.8, the true area of contact by one asperity is,

$$A_i = \pi a^2 = \lambda\pi R\delta \tag{3.22}$$

where λ is an area factor which depends on the degree of work hardening of the metal.

For elastic contact, i.e. at $p = 1$, $\lambda = 1$

For plastic mode, i.e. at $p = o$, $\lambda = 2$

From equations 3.19 and 3.22, the load supported by one asperity is

$$W_i = K\lambda^{\left(1 + \frac{p}{2}\right)} R^{\left(1 - \frac{p}{2}\right)} \delta^{\left(1 + \frac{p}{2}\right)}$$

The compliance is (z-d). Therefore

$$A_i = \lambda\pi R(z-d)$$

and

$$W_i = K\lambda^{\left(1 + \frac{p}{2}\right)} R^{\left(1 - \frac{p}{2}\right)} (z-d)^{\left(1 + \frac{p}{2}\right)}$$

The probability of making contact at any asperity of height z above a datum line is

$$\text{prob } (z>d) = \int_d^\infty \phi(z)dz$$

If there are n asperities per unit area, the total true area of contact A_λ is

$$A_\lambda = \lambda\pi nR \int_d^\infty (z-d)\phi(z)dz$$

The total load W_λ is

$$W\lambda = nK\lambda^{\left(1 + \frac{p}{2}\right)} R^{\left(1 - \frac{p}{2}\right)} \int_d^\infty (z-d)^{\left(1 + \frac{p}{2}\right)} \phi(z)dz$$

As in previous sections for elastic interaction, if A is the apparent area,

$$A_\lambda = \lambda\pi nAR\sigma F_1(h)$$

and

$$W_\lambda = nAK\lambda^{\left(1+\frac{p}{2}\right)} R^{\left(1-\frac{p}{2}\right)} \sigma^{\left(1+\frac{p}{2}\right)} F_{\left(1+\frac{p}{2}\right)}(h)$$

where, as before

$$F_m = \int_h^\infty (s-h)^m \phi^*(s)ds$$

considering the upper decile of a real surface (Fig. 3.4) to have an exponential distribution,

$$\frac{W_\lambda}{A_\lambda} = \frac{K}{\pi} \left(\frac{\lambda\sigma}{R}\right)^{\frac{p}{2}} \left(1 + \frac{p}{2}\right) \; ; \qquad (3.23)$$

Equation 3.23 shows that, for a given surface which has work hardened, the relationship between the true area of contact and the applied load is linear. If the encounter is elastic when p = 1, the area of contact is expressed in terms of the peak height distribution of the asperity summits. For plastic deformation at p = o, the true area of contact is a function of π/K, which can be identified as being the flow stress of the metal as is seen in equation 3.19.

The statistical treatment discussed here shows that the contact of solids is controlled by three topographic parameters viz., the surface density of the asperities, their mean radius and the standard deviation of the peak height distribution together with the mechanical properties of the metals, which has given the concept of the plasticity index (chapter 14). The normal load determines the number of individual asperity contacts and an increase in the load causes the compliance to change as the surfaces move together. The prior areas of contact increase in size but this is balanced by the formation of smaller areas. In other words, although both the number and the total area of true contact are directly proportional to the load, individual contacts fall in a range of sizes and the mean contact width is not significantly influenced by the normal load.

REFERENCES

1. Archard J F, *Proc Roy Soc A* (1957) *243*, 190.
2. Halling J (Ed), *'Principles of Tribology'*, The Macmillan Press (1975).
3. Greenwood J A and Williamson J B P, *Proc Roy Soc A*, (1966) *295*, 300.
4. Halling J and Nuri K A, *Proc IUTAM Symposium on Contact Mechanics, Holland* (1974).

CHAPTER 4

FRICTION

When one surface is placed upon another, a finite amount of horizontal force is necessary to initiate sliding. This horizontal force is the static friction between the two surfaces, the magnitude of which depends, inter alia, on the cleanliness of the interface so that friction between two bodies is high in vacuum. The force necessary to maintain sliding must be greater than the kinetic friction between the two surfaces. The resistance to slide under tangential traction has for a long time been explained by the roughness hypothesis which suggests that the peaks of one surface rest in the valleys of the other and the resistance to motion is the effort expended by the former to climb out of the latter. The current theory of friction starts with the basic premise that contact occurs at a few isolated points only because of the undulating nature of surfaces. The contact areas form metallic junctions because of plastic or elastic deformation so that frictional resistance is the force necessary to shear these junctions.

4.1 *Area of Contact*

Consider a surface loaded against another (Fig. 4.1), both having protuberances as is the case with all engineering surfaces. Contact can not occur over the whole apparent area of the interface and the two solids can only meet at those asperities which have approached one another favourably. This way areas of contact a_1, a_2, etc will form giving the true area of contact A_t such that

$$A_t = a_1 + a_2 + a_3 + \dots + a_n \tag{4.1}$$

Measurement of electrical conductivity of stationary surfaces shows[1] that the true contact area is a mere fraction of the apparent area of contact. For example, for steel surfaces, $A_t = (A/10\,000)$, where A is the apparent area of contact which is the measured surface area of the interface. A small area of contact means that the load is supported at a few isolated points. This results in an intensified normal stress at the interface and it is well established that the material at a_1, a_2, a_3 . . . . a_n flows plastically and forms strong junctions between surfaces 1 and 2 (Fig. 4.1). Obviously, the sum of the interfacial areas of all the junctions is the true area of contact and these must be broken in order that one surface can slide relative to the other.

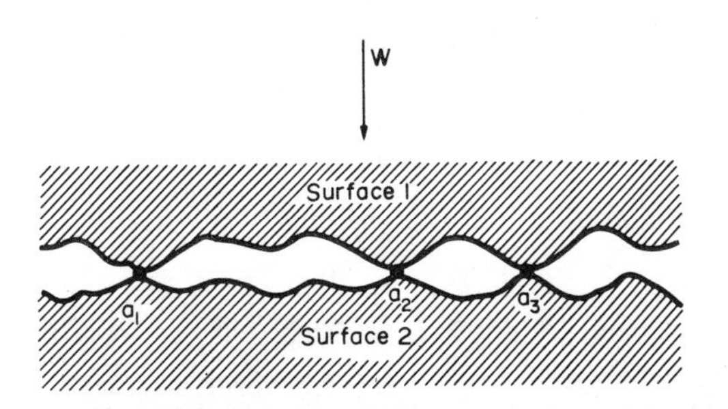

Fig. 4.1 A surface 1 resting on a surface 2 under a normal load W. Contact occurs only at a few asperities, a_1, a_2 and a_3.

4.2 *Adhesion of Junctions*

Development of a finite strength at the interface by plastic interaction depends largely on the mechanical properties of the material. Thus if the applied normal load on surface 1 (Fig. 4.1) is W and σ_y is the flow pressure of the softer element of the two, for an established true contact area A_t,

$$W = A_t \sigma_y$$

$$\text{or} \qquad A_t = \frac{W}{\sigma_y} \tag{4.2}$$

If the load is removed, the surfaces may separate if the junctions ruptured by the process of elastic recovery of the underlying material[2]. Thus, consider a hard spherical indenter of radius r_1 and Young's modulus E_1 pressing under a load W on a soft flat surface of Young's modulus E_2 making a maximum depth of indentation x (Fig. 4.2a). Let the chordal diameter of the indentation be d. Suppose that the load is released and, as a result of elastic recovery, the softer surface springs back such that it has a new radius of curvature r_2, $r_2 > r_1$, and it can often be assumed that the chordal diameter is the same as before.

The amount of elastic recovery can be assessed by using an arbitrary reference point on the sphere and noting the height by which it has moved up from the original surface once the load is removed. Let the elastic recovery, as defined, be h and let y be the distance from the original surface to the horizontal line drawn through the centre of the final surface of radius r_2 (Fig. 4.2b). It follows that

$$h = x - y \tag{4.3}$$

Now from Fig. 4.2a,

$$(2r_1 - x)x = \frac{d}{2} \times \frac{d}{2}$$

$$\text{or} \qquad 2r_1x - x^2 = \frac{d^2}{4}$$

Since x is usually small, the higher powers of x can be neglected, so that

$$2r_1x = \frac{d^2}{4}$$

or

$$x = \frac{d^2}{8r_1} \tag{4.4}$$

Similarly, from Fig. 4.2b,

$$(2r_2 - y)y = \frac{d^2}{4}$$

or

$$y = \frac{d^2}{8r_2} \tag{4.5}$$

therefore

$$h = (x - y)$$

$$= \frac{d^2}{8}\left(\frac{1}{r_1} - \frac{1}{r_2}\right), \text{(from equations 4.3, 4.4, and 4.5)}$$

or

$$h = \frac{d^2}{8}\left(\frac{r_2 - r_1}{r_1 r_2}\right) \tag{4.6}$$

Now defining E as

$$\frac{1}{E} = \frac{1 - \upsilon_1^2}{E_1} + \frac{1 - \upsilon_2^2}{E_2}$$

where υ_1 and υ_2 are the respective poisson's ratios

and taking $\upsilon_1 = \upsilon_2 = 0.3$

$$\frac{1}{E} = \frac{0.91\ (E_1 + E_2)}{E_1E_2}$$

or

$$E = \frac{E_1E_2}{0.91\ (E_1 + E_2)}$$

Defining r as

$$\frac{1}{r} = \frac{1}{r_1} - \frac{1}{r_2}$$

i.e. $r = \dfrac{r_1 r_2}{r_2 - r_1}$

and substituting the values of E and r in equation 3.4

$$d = 2\left[\frac{3}{4} W \frac{r_1 r_2}{(r_2 - r_1)} \quad \frac{0.91\ (E_1 + E_2)}{E_1 E_2}\right]^{\frac{1}{3}}$$

$$\text{or } d = \left[5.46\ W \left(\frac{r_1 r_2}{r_2 - r_1}\right) \left(\frac{1}{E_1} + \frac{1}{E_2}\right)\right]^{\frac{1}{3}}$$

substituting for $(r_1 r_2)/(r_2 - r_1)$ from equation 4.6

$$d^3 = \left[5.46\ W \left(\frac{d^2}{8h}\right) \left(\frac{1}{E_1} + \frac{1}{E_2}\right)\right]$$

$$\text{or } h = 0.68 \frac{W}{d} \left(\frac{1}{E_1} + \frac{1}{E_2}\right)$$

And multiplying the numerator and denominator of the right hand side by d,

$$h = 0.68d\ \frac{W}{d^2} \left(\frac{1}{E_1} + \frac{1}{E_2}\right)$$

Assuming that for a given load W, the indentation diameter d is constant for any material,

$$h = K\left(\frac{1}{E_1} + \frac{1}{E_2}\right) \qquad (4.7)$$

where $K = 0.68d \left(\dfrac{W}{d^2}\right) = \text{constant}$

The ratio W/d^2 is a measure of the hardness of the indented metal so that equation 4.7 shows that the harder the indented metal the greater is its elastic recovery. Alternatively, as the Young's modulus of the indented metal is increased, spring back of the rider becomes high. Calculation shows that for a hard steel rider indenting on mild steel and indium respectively, the value of h (Fig. 4.2) for steel can be 15 times that of indium. Since the integrity of the junction decides the magnitude of frictional resistance, steel surfaces should give a lower frictional force than those of indium when the indenter is the same for both metals.

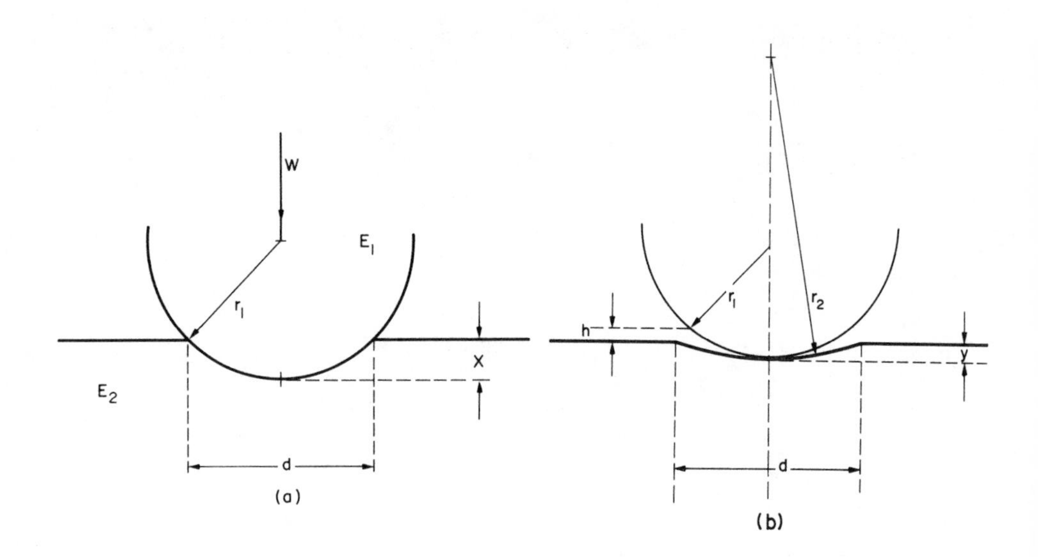

Fig. 4.2 Indentation of a soft flat surface by a hard sphere. (a) a load W causes an indentation of diameter d and depth x. (b) The load is removed causing the surface to spring back to a depth $y < x$ and it is assumed that the indentation diameter d does not change. (McFarlane and Tabor[2]).

4.3 *Mechanism of Friction*

The pioneering work providing evidence of the true contact area being a mere fraction of the apparent area should be attributed to Holm[4] but it would appear that the mechanism of friction embodying adhesion gained acceptance subsequent to the work of Hardy and Hardy[5] reported in 1919 and again a decade later to the report by Tomlinson[6]. Some basic postulates of Tomlinson have been criticised and his work is summarised in chapter 6. If the reader is interested in the chronological order, he should read chapter 6 now and then proceed with the material presented here.

Bowden and Tabor[7] consider the frictional resistance between two surfaces as the sum of a shearing and a ploughing component. As in the previous section a hard hemispherical rider of radius r is loaded against a flat surface[8]. Suppose a normal load W makes the flat surface to form a cup AOB whose maximum height OC is h and whose chord diameter is d (Fig. 4.3). Let the frictional resistance be F which is the minimum tangential force necessary to initiate sliding.

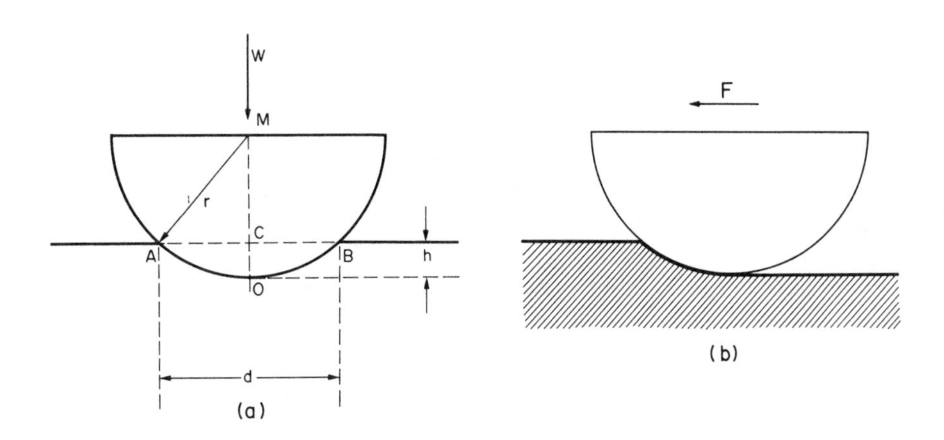

Fig. 4.3 A hemispherical rider under load sliding on a soft surface. (a) rider stationary (b) rider moving, producing a grooved track. (Bowden, Moore and Tabor[8])

The effect of the load W is to cause yielding of those asperities of the flat surface which have made contact with the hard rider. That is, the surfaces of the two bodies have adhered wherever contact occurred. This gives rise to the shearing term which is that the junctions so formed by plastic yielding must be broken completely before the rider can be made to move over the soft surface. If τ is the shear strength of the softer metal which has flowed plastically and A_t is the true area of contact at the interface, the shearing term S is

$$S = A_t\tau \tag{4.8}$$

It is evident that for the rider to move horizontally, an additional force is also required to displace a wall of metal ahead of the hemisphere in the direction of motion as it sinks into the softer metal. This is the ploughing term P and is given by

$$P = A'\sigma_y \tag{4.9}$$

where A' is the area of the grooved track and can be regarded as that of the segment AOBC (Fig. 4.3a). To a first approximation, the area of the segment AOBC is

$$\frac{h(4d^2 + 3h^2)}{6d} = \frac{2hd}{3},$$ neglecting the higher powers of h, since h is small.

Now from Fig. 4.3a, for the right angled triangle MAC,

$$(r - h)^2 + \frac{d^2}{4} = r^2$$

and neglecting h^2,

$$h = \frac{d^2}{8r}$$

substituting this value of h in the area of the segment,

$$\text{Area AOBC} = \frac{d^3}{12r}$$

Therefore the frictional resistance due to the ploughing term is, from equation 4.9

$$P = \frac{d^3}{12r} \sigma_y \qquad (4.10)$$

Since the total frictional force F is given by the sum of the ploughing and the shearing term,

$$F = S + P$$

substituting for S and P from equations 4.8 and 4.10 respectively,

$$F = A_t\tau + \frac{d^3}{12r} \sigma_y \qquad (4.11)$$

For hard metals loaded against similar surfaces or, possibly, for most combinations at light loads, the depth of indentation is small so that the term d in equation 4.10 can be neglected and the ploughing term becomes insignificant. In that case,

$$F = A_t\tau \qquad (4.12)$$

The coefficient of friction μ is the frictional resistance per unit load. Thus,

$$\mu = F/W = \frac{A_t\tau}{A_t\sigma_y}$$

or

$$\mu = \tau/\sigma_y \qquad (4.13)$$

4.4 *Amontons' Laws*

The preceding discussions confirm the laws enunciated by Amontons in 1699 which are as follows:

First Law : The frictional force is independent of the apparent area of contact and is proportional to the applied load only.

Second Law: The coefficient of friction is independent of the applied load.

REFERENCES

1. Bowden F P and Tabor D, *Proc Roy Soc A*, (1939), *169*, 391.
2. McFarlane J S and Tabor D, *Proc Roy Soc A*, (1950), *202*, 224.
3. Whitehead J R, *Proc Roy Soc A*, (1950), *201*, 109.
4. Holm R, *'Electrical Contacts'*, H Gerbers, Stockholm, (1946).
5. Hardy W B and Hardy J K, *Phil Mag*, (1919), *38*, 32.
6. Tomlinson G A, *Phil Mag*, (1929), *7*, 905.
7. Bowden F P and Tabor D, *'The Friction and Lubrication of Solids'*, Part 1, Oxford University Press, (1964).
8. Bowden F P, Moore A J W and Tabor D, *Jl Appl Physics*, (1943), *14*, 80.

CHAPTER 5

EFFECT OF SLIDING

Since a metal couple under load will form junctions by plastic flow of favourably disposed surface peaks, the first effect of sliding is to rupture these and effect a change in surface topography. Sliding inevitably generates frictional heat which is responsible for two important metallurgical effects. Firstly, there is thermal softening of the interface with a concomitant lowering of yield stress of the metals. Secondly, unless the operation is carried out in vacuum, oxidation of the interface is facilitated. A low yield stress means that asperity yielding becomes easier but the presence of a partitioning film of oxide dilutes the metal interaction and hence lowers the frictional resistance.

5.1 *Junction Growth*

Equation 4.13 showed that the coefficient of friction $\mu = \tau/\sigma_y$, where τ is the shear strength of the junction and σ_y its flow pressure. Shear strength of such junctions can only be obtained by dividing the measured frictional resistance with the true area of contact which is not very easy to establish. It is a fair assumption, however, to suggest that $\tau = \tau_c$, where τ_c is the critical shear stress of the softer member of the couple. The implication is, of course, that the strength of the interface is solely dictated by the adhesion of the soft metal on to the hard surface. Since $\sigma_y \simeq 5\tau_c$, $\mu = \frac{\tau}{\sigma_y} \simeq \frac{\tau}{5\tau_c} = 0.2$. Bowden and Tabor[1] point out that for most dry sliding situations in room atmosphere, $\mu \simeq 1$. One could argue that the observed values of μ are high because equation 4.13 ignores work hardening which would cause an increase in the shear strength of the interface. However, it should be remembered that the effect of work hardening is to cause an increase in the value of the flow pressure also.

The discrepancy between the observed and calculated values of coefficient of friction has been explained in terms of junction growth, that is, a lateral growth of the true contact area upon the application of a tangential pull. Work of McFarlane and Tabor[2] who slid steel and indium, a material with a low work hardening characteristic, and that of Parker and Hatch[3] who used lead show that the initial effect of a tangential pull is to cause microdisplacement of the contact spots, generally in the horizontal direction. The microdisplacement of junctions under the combined action of normal and tangential loading has been shown[4] to occur even before macroscopic sliding on a large scale takes place.

Applying Von Mises' criterion, for an interface under a normal stress σ

$$\sigma^2 + \alpha\tau^2 = \sigma_y^2 \tag{5.1}$$

where α is a constant.

Equation 5.1 suggests that if the true contact area increases, σ must decrease and, therefore, τ must increase. This should explain the reason for observed high values of friction.

5.1.1 *Equation for Junction Growth.* The basic equation for junction growth is given[1] as

$$\sigma^2 + \alpha_o \tau_t^2 = \alpha_o \tau_m^2 \tag{5.2}$$

where α_o is a constant with a value of about 9. τ_t is the shear stress at the interface or, more specifically, the applied tangential stress to cause rupture of the junctions and τ_m is the critical shear stress of the soft metal. Equation 5.2 suggests that as the tangential force is increased monotonically from zero up to the value when sliding is just about to commence, $\sigma \to o$ because of microdisplacement of the contact area and consequently $\tau_t \to \tau_m$.

5.2 *Work of Adhesion*

The foregoing discussion suggests that the interfacial shear stress of two interacting bodies can never be greater than the critical shear stress of the junctions. Tamai[5] finds equation 4.13 inadequate on the grounds that the tangential force to cause and maintain sliding varies between different materials although they may possess comparable mechanical properties.

Machlin and Yankee[6] postulate that the magnitude of the frictional force is probably decided in accordance with the ability for junction growth by the solid phase weldability of a couple, which may be defined as the ratio of the work of adhesion of the surfaces to the strength of the weaker component.

Rabinowicz[7, 8] uses the work of adhesion E between two interacting surfaces to explain friction and shows that this gives a better agreement than equation 4.13 between the theoretical and experimental values for the coefficient of friction. The work of adhesion between two dissimilar metals a and b is given by

$$E_{ab} = G_a + G_b - G_{ab} \tag{5.3}$$

and for like metal couples, e.g. metal a,

$$E_a = 2G_a \tag{5.4}$$

where G_a and G_b are the surface free energies of metals a and b respectively and G_{ab} is the interfacial free energy when the two surfaces make contact.

Consider (Fig. 5.1) a hard conical asperity under a load ΔW making an indentation of depth x in a soft surface. Let the indented diameter be 2r and θ be the base angle of the asperity.

Neglecting any elastic component, that is considering the interaction to be wholly plastic, the following energy terms are involved.

(i) The plastic energy of deformation of the soft metal and this is given by $\int_o^x \pi r^2 \sigma_y dx$, where σ_y is the flow pressure of the soft metal.

(ii) Work done by the load ΔW to move a distance x, which equals ΔWx.

(iii) The formation of a true contact area results in a lowering of the surface energy of the system and the magnitude of this is $E_{ab}\pi r^2$

Thus the energy balance of the system is

$$G - E_{ab}\pi r^2 = \Delta Wx - \int_0^x \pi r^2 \sigma_y dx \tag{5.5}$$

where G is the total energy in the system.

From Fig. 5.1, putting $r = x \cot \theta$

$$G = E_{ab}\pi x^2 \cot^2\theta + \Delta Wx - \pi\sigma_y \cot^2\theta \int_0^x x^2 dx$$

Integrating the last term,

$$G = E_{ab}\pi x^2 \cot^2\theta + \Delta Wx - \frac{\pi\sigma_y \cot^2\theta x^3}{3} \tag{5.6}$$

At equilibrium, that is when the process of indenting has ceased,

$\frac{dG}{dx} = o$. Hence differentiating equation 5.6,

$$\frac{dG}{dx} = 2\pi E_{ab} x \cot^2\theta + \Delta W - \pi\sigma_y x^2 \cot^2\theta$$

Putting $x \cot\theta = r$, at equilibrium, i.e., at $\frac{dG}{dx} = o$,

$$\Delta W = \pi r^2 \sigma_y - 2\pi r E_{ab} \cot\theta$$

Dividing both sides by πr^2

$$\frac{\Delta W}{\pi r^2} = \sigma_y - \frac{2E_{ab}\cot\theta}{r} \tag{5.7}$$

Now neglecting the ploughing term due to a tangential pull, the coefficient of friction μ for the asperity in Fig. 5.1 is

$\mu = \frac{\tau\pi r^2}{\Delta W}$, where τ is the shear stress at the interface.

Thus equation 5.7 can be written as

$$\frac{\tau}{\mu} = \sigma_y - \frac{2E_{ab}\cot\theta}{r}$$

or

$$\mu = \frac{\tau}{\sigma_y - \left(\dfrac{2E_{ab}\cot\theta}{r}\right)} \qquad (5.8)$$

Equation 5.8 shows that $\mu \to \infty$, when

$$\sigma_y = \frac{2E_{ab}\cot\theta}{r}$$

i.e.

$$\frac{E_{ab}}{\sigma_y} = \frac{r}{2\cot\theta}$$

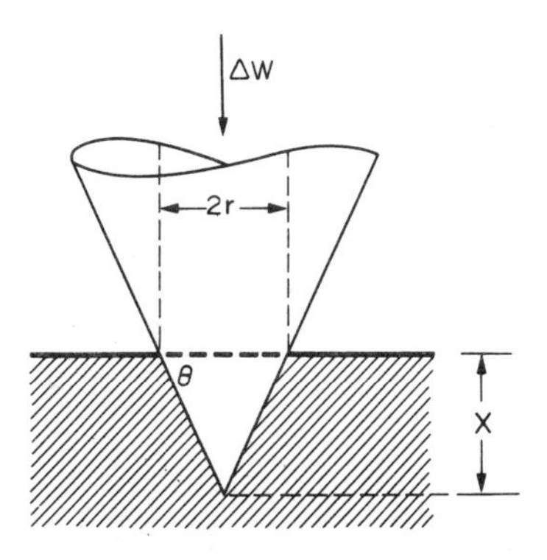

Fig. 5.1 A conical hard asperity indenting a soft surface.

5.3 *Kinetic Friction*

Both equations 4.13 and 5.8 show a dependence of the frictional force on the shear strength of the adhered surfaces. A junction forms and grows radially so that the area of true contact assumes a large enough size to support the normal load. The area spreads out further in the direction of motion as sliding commences. Having regard to equation 5.1, the normal stress should decrease with a concomitant rise in the shear stress of the interface. That is, with the onset of a tangential pull, a maximum static frictional resistance will develop until, with increasing external force, the interface will separate. At that instant frictional resistance is zero but fresh junctions form elsewhere and the process is continued this way. Rabinowicz[9] shows that, depending on the nature of the surfaces, the static coefficient of friction persists for a distance of the order of 10^{-4} cm as sliding commences and its magnitude then falls up to a distance of 10^{-3} cm when the kinetic component of frictional resistance is reached. This is shown schematically in Fig. 5.2.

Consolidation of the junctions and hence the magnitude of the resistance to sliding depends on the life of the stationary contact, being small when the contact time is measurable in milliseconds and large when this is a few seconds. In a typical tribological situation, the contact time between two sets of interacting asperities is small and, hence, the kinetic friction should always be lower than the static component which is the force necessary to initiate sliding between two initially stationary surfaces. An increased strength of a static interface under load with time is attributed to a process analogous to creep when the junctions grow in all directions.

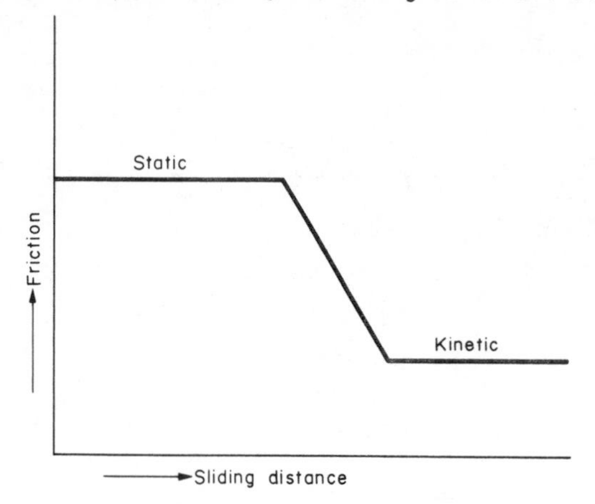

Fig. 5.2 Schematic representation of static and kinetic friction between two surfaces with the onset of sliding. Static friction persists at the beginning of motion but soon falls to the value of the kinetic friction.

5.4 *Stick-Slip*

It follows that, in a sliding situation, a smooth motion giving a steady friction should not be expected. As the asperities adhere during an encounter, the moving parts of a machine should stick followed by a high value of friction. If the external force is maintained and the junctions rupture, the system will slip and the friction will tend to zero. Adhesion of asperities followed by the rupture of the interface, that is the process of stick-slip inevitably occurs as dissimilar metals slide under load. Similar metal couples also show large fluctuations in friction but these are comparatively slow and very irregular.[10]

Frictional effects extend beyond the surface to a depth below into the bulk material so that the mechanical properties of an interacting couple would also contribute to the nature of the interfacial forces. Experiments with various metals sliding on steel have shown a dependence of slip on the melting points of the materials.

The nature of the friction traces obtained by Bowden et al[10, 11] is shown schematically in Fig. 5.3. With the onset of motion the magnitude of friction increased as shown by AB. It was observed that, during this period, the rider moved forward with the same velocity as the slider. That is, in the time interval A to B the surfaces were sticking. At B there was rapid slip between the surfaces and the frictional force diminished to C. Simultaneous measurement of surface temperature showed that every time a slip movement occurred there was a sudden rise and fall in temperature, the whole thermal flash lasting for less than a thousandth of a second. This temperature flash and the stick-slip movement continued throughout the experiment.

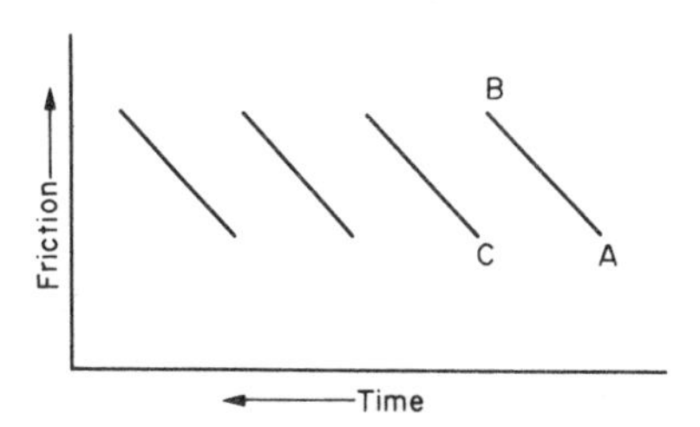

Fig. 5.3 Stick-slip phenomena (after Bowden, Leben and Tabor[11]).

The high melting point metals, such as molybdenum, cut a grooved track in the steel, the width of which was seen to be the larger, the higher the melting point of the rider. Low melting point metals such as zinc formed a smeared layer on the steel surface. Aluminium showed a transitional behaviour in that it both scratched the surface and deposited a layer of metal.

From detailed experiments, the authors conclude that there are three types of sliding as follows.

(1) When a hard high melting point rider slides on a soft lower melting point metal, the motion is jerky but sudden slip is not so obvious (Fig. 5.4a). The rider remains unaffected, but a track is torn in the softer metal. It is suggested that frictional resistance is due mainly to the ploughing of the softer metal.

(2) When a low melting point metal slides on a hard surface, the motion is jerky and the extent of the rapid slip is much greater (Fig. 5.4b) than that in the first type (Fig. 5.4a). The hard surface is not scratched and the soft metal is smeared over it. It is suggested that high contact pressure and frictional heat liberated during rapid slip solders the low melting point metal on to the hard slider. As the surfaces begin to stick, the solders are drawn out thinly until during the next slip when soldering occurs again.

(3) For similar metals, frictional forces are very much higher than those for the last two combinations and large fluctuations are observed (Fig. 5.4c) without any rapid slip. A large groove forms in the flat surface and welding and incipient fusion again occurs causing damage to both surfaces.

5.5 *Thermal Effect*

In 1903, Beilby[12] reported his observation that polishing of metals produced enough frictional heat to result in a surface layer which was physically different from the bulk of the material. Presence of this Beilby layer, as it was named, was confirmed by Cochrane[13] in 1938 from an analysis of the electron diffraction patterns of polished surfaces. Such surfaces are now known to possess a high dislocation density, giving a disordered structure and is technologically important as an engine is being run-in because this layer is harder and tougher than the underlying metal[14]. Formation of such a surface layer due to polishing has also been confirmed by Bowden and Hughes[15].

Bowden and Ridler[16] show a linear increase in temperature of sliding surfaces with speed or load until the melting temperature of one of the elements

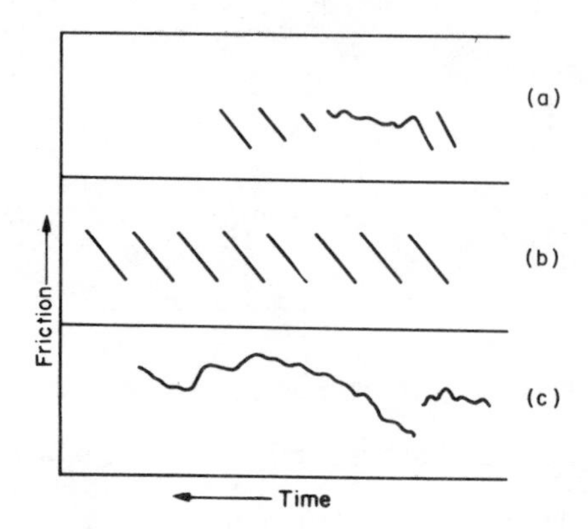

Fig. 5.4 Stick-slip of three different metal combinations. (schematic, after Bowden, Leben and Tabor[11]) (a) Hard pin on soft surfaces; (b) Soft pin on hard surface; (c) Similar metals.

forming the couple is reached when the temperature remains at a constant value. Their experimental set up comprised a rotating annular ring upon which a pin was loaded and both the ring and the pin were connected to form a thermocouple. Some typical results obtained by sliding gallium on mild steel is shown in Fig. 5.5 where, on the ordinate, T is the recorded temperature and T_0 is the ambient temperature. That is, $T - T_0$ is the rise in the bulk temperature of the interface as a result of frictional heat generated during sliding.

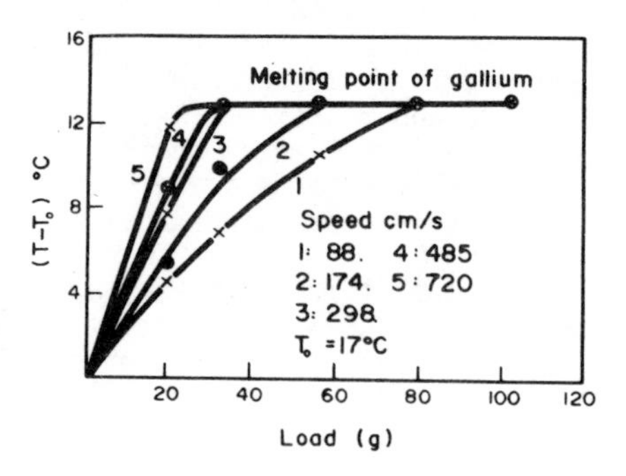

Fig. 5.5 Variation of temperature with load for gallium sliding on mild steel (Bowden and Ridler[16]).

It is known that there is usually a deterioration in mechanical properties of metals as the temperature of the surrounding environment is raised. In consequence, an increase in surface temperature should facilitate plastic flow of the asperities under normal and tangential stresses. If the contact area spreads, the normal stress σ in equation 5.1 should decrease, giving a high value of τ and hence of the frictional resistance. However, since the metal is weak at high temperature, the critical shear strength of the junctions should fall and a low value of the coefficient of friction is expected in spite of an increase in the contact area. This is true for most metal combinations with some exceptions. For example, Bowden and Hughes[17] found that, upon sliding gold on gold, the coefficient of friction rose to a value of the order of 30 at an experimental temperature of about 600°C. The authors[17] suggest that, since gold is a soft metal, the contact area increases readily resulting in a high frictional resistance. This, unfortunately, poses the embarrassing anomaly that the shear strength of the junctions decreases with a rise in surface temperature.

These experiments bring out an interesting point which is that calculations[16] show that for a 1 mm diameter constantan cylinder sliding at 100 cm/s on mild steel at a normal load of 100 g, the rise in interface temperature is 75°C assuming that the whole surface is in contact. However, if the true contact area is $\frac{1}{10}$ th of the apparent area, the temperature of the junctions is 414°C and it is 2372°C when the fraction is taken as $\frac{1}{100}$. It is permissible to speculate if the rise in friction between metals at high temperature is as a result of the highly viscous molten metal or, conversely, a fall in friction is due to the lubricating effect of a fluid film of metal at the interface.

5.6 *Oxide Film*

The theories of friction so far discussed assume that metal junctions are chemically clean before and during interaction, that is they are free from contaminants. It is possible to clean metals free of grease etc. but it is an ubiquitous feature that most metals are covered with a layer or more of an oxide in room atmosphere. The effect is that metal to metal interaction is diluted by a non-plastic medium at the sliding interface. Heavy loads may be capable of breaking up the coherency of the oxides but, at light loads[18], with metals which can form strong oxides, the frictional resistance is low.

The efficiency of the role of oxidation is controlled by surface preparation, an abraded surface giving lower friction than the one prepared electrolytically. Experiments[19] with a number of metals within a load range of 0.0039 to 10 kg show that presence of oxides does not necessarily mean that frictional forces are low but that there is less surface damage. This contradicts the postulate that for the frictional resistance to be high, strong metallic junctions must form giving rough surfaces as they rupture. However, hard metals tend to assume a polished appearance as they slide. They give low values of friction generally[20] and it is known[21] that they undergo considerable plastic interaction.

Experiments[22] in a vacuum of 10^{-6} torr show that couples give a high value of friction but this falls quickly in about 2 seconds if an amount of oxygen is introduced into the system. As the couples are given prolonged exposure in the oxygen atmosphere, it would appear that an oxide film forms and thickens to a constant value when the coefficient of friction also settles down at a constant figure. Tingle[23] suggests, however, that it is the oxide film immediately adjacent to the metal surface that decides on the magnitude of the frictional resistance and not its thickness which increases with time.

5.7 *Sliding between Brittle Surfaces*

An oxide is largely brittle and interaction of surface asperities of brittle materials should not give rise to plastic flow, although it has been observed[14] to happen with certain non-metals. In the absence of plastic flow, adhesion of surfaces and the subsequent shearing of junctions would seem unlikely. Byerlee[24] proposes a model where, at an encounter, the asperities fail by brittle tensile fracture rather than by plastic shear. Measured values of the coefficient of friction with materials such as quartz agree well with calculated values and the coefficient of friction is found to increase as the surface becomes rougher.

5.8 *Effect of Contaminants on Friction*

Real surfaces are contaminated either or both with oxides or atmospheric gases and these are discussed in detail later in chapters 9 and 10. Contaminants, if non-corrosive, are beneficial from the view point of friction and wear and in this section a simple expression for the coefficient of friction for contaminated surface is deduced, which finds wide application.

An interesting phenomenon observed in experiments is that, at any load, the true area of contact is the same for both lubricated and unlubricated conditions. Growth of the junctions occurs in both cases as a tangential pull is applied although the lubricated junctions cease to grow sooner than the clean areas of contact.

Tabor[25] suggests that a contaminant forms an interfacial layer between the two surfaces whose critical shear stress τ_i is less than τ_m, the critical shear stress of the metal junctions. As long as the applied tangential stress τ_t does not exceed τ_i, the normal and tangential stresses can be transmitted through this interfacial layer to the metal substrate for it to flow plastically according to equation 5.2. Growth of the points of contact occurs and, when $\tau_t = \tau_i$, sliding commences as the interfacial layer shears.

If the true area of contact at this stage is A_i, the frictional force F is

$$F = A_i \tau_i \tag{5.9}$$

Assume now that

$$\tau_i = f \tau_m \tag{5.10}$$

where f is less than unity.

Since sliding has commenced, $\tau_t = \tau_i$ and substituting for τ_m from equation 5.10 in equation 5.2,

$$\sigma^2 + \alpha_o \tau_i^2 = \alpha_o \left(\frac{\tau_i}{f} \right)^2$$

or

$$\sigma^2 = \alpha_o \left(\frac{\tau_i^2}{f^2} - \tau_i^2 \right)$$

or

$$\sigma^2 = \alpha_o \tau_i^2 \left(\frac{1 - f^2}{f^2} \right) \tag{5.11}$$

Now $\mu = \frac{F}{W}$ where W = applied normal load.

and $W = A_i \sigma$

Substituting for F from equation 5.9,

$$\mu = \frac{A_i \tau_i}{A_i \sigma} = \frac{\tau_i}{\sigma}$$

Substituting for σ from equation 5.11

$$\mu = \frac{f}{\left\{\alpha_0 (1-f^2)\right\}^{\frac{1}{2}}} \qquad (5.12)$$

For metal surfaces free from any contaminants, f = 1 and equation 5.12 shows that the coefficient of friction is then very high. For calculations, α_0 is taken as 9.

REFERENCES

1. Bowden F P and Tabor D, *'The Friction and Lubrication of Solids'*, Part 2, Oxford University Press, (1964).
2. McFarlane J S and Tabor D, *Proc Roy Soc A*, (1950), *202*, 244.
3. Parker R C and Hatch D, *Proc Phys Soc B*, (1950), *63*, 185.
4. Courtney Pratt J S and Eisner E, *Proc Roy Soc A*, (1957), *238*, 529.
5. Tamai Y, *Jl Appl Physics*, (1961), *32*, 1437.
6. Machlin E S and Yankee W. R, *Jl Appl Physics*, (1954), *25*, 576.
7. Rabinowicz E, *Jl Appl Physics*, (1961) *32*, 1440.
8. Rabinowicz E, *Trans ASLE*, (1958), *1*, 96.
9. Rabinowicz E, *Jl Appl Physics*, (1951), *22*, 1373.
10. Bowden F P and Leben L, *Proc Roy Soc A*, (1939), *169*, 371.
11. Bowden F P, Leben L and Tabor D, *The Engineer (London)*, (1939), *168*, 214.
12. Beilby G T, *Proc Roy Soc A*, (1903), *72*, 218.
13. Cochrane W, *Proc Roy Soc A*, (1938), *166*, 228.
14. Finch G I, *Science Progress*, (1937), *31*, 609.
15. Bowden F P and Hughes T P, *Proc Roy Soc A*, (1937), *160*, 575.
16. Bowden F P and Ridler K E W, *Proc Roy Soc A*, (1936), *154*, 640.
17. Bowden F P and Hughes T P, *Proc Roy Soc A*, (1939), *172*, 263.
18. Whitehead J R, *Proc Roy Soc A*, (1950), *201*, 109.
19. Wilson R W, *Proc Roy Soc A*, (1952), *212*, 450.
20. Moore A J W and McG.Tegart W J, *Proc Roy Soc A*, (1952), *212*, 452.
21. Rabinowicz E and Tabor D, *Proc Roy Soc A*, (1951), *208*, 455.
22. Bowden F P and Young J E, *Proc Roy Soc A*, (1951), *208*, 311.
23. Tingle E D, *Trans Faraday Soc*, (1950), *46*, 93.
24. Byerlee J D, *Jl Appl Physics*, (1967), *38*, 2928.
25. Tabor D, *Proc Roy Soc A*, (1959), *251*, 378.

CHAPTER 6

MOLECULAR THEORY OF FRICTION AND WEAR

The molecular theory of friction and wear was propounded by Tomlinson[1] in 1929 who considered at length the nature of atomic forces in a crystal lattice and deduced expressions for dry friction and wear of solids. In what follows, the situations, where dry surfaces are under continued unidirectional motions, are presented, although Tomlinson has also discussed rolling friction. Students should study this chapter carefully and then compare critically the more recent theories given in chapter 5. Later on (chapter 14), the concept of plasticity index is introduced. This will allow an opportunity to re-examine the ideas of friction given in chapter 5.

6.1 *Dry Friction*

Essentially, Tomlinson's hypothesis is that, at equilibrium, the repulsive forces between atoms in a solid counteract the cohesive forces. However, when two bodies are in contact, an atom of one body may come sufficiently close to that of the second to enter the field of repulsion. When this happens, the two surfaces separate causing a loss of energy which manifests itself as the resistance due to friction.

Tomlinson recognises that engineering surfaces are rough so that not all the surface atoms of the two interacting bodies will react. In other words, some of them will be too far apart to enter the field of repulsion, that is at a given load a number of atomic contacts n_0 will exist at the interface under static condition. The total force P between the two surfaces is assumed to be supported by the sum of the atomic forces of repulsion, p. Thus

$$P = \Sigma(p) \tag{6.1}$$

However, the magnitude of the force of repulsion at each junction should vary since the two bodies are not expected to fit together exactly so that the distances between the approaching atoms will be different. It is necessary, therefore, to take an arithmetic mean value of the individual forces of repulsion. Denoting this mean repulsive force by p_0

$$P = n_0 p_0 \tag{6.2}$$

If now one surface slides a distance x relative to the other, the mechanical work done is μPx, where μ is the coefficient of friction. In the process of sliding, new atoms will enter the field of repulsion and others will leave it. There will thus be a total loss of energy, the value of which will be given by the sum of the energies lost at each encounter for that particular amount of sliding which has occurred. If E is the arithmetic mean value of all the energies lost due to atom to atom collision and if the number of encounters for a distance x is n, the total energy lost is nE, which is equal to the mechanical work done. That is,

$$nE = \mu Px \tag{6.3}$$

An expression for n in terms of the interatomic distance e of the crystal is obtained as follows.

Consider a static interface (Fig. 6.1) AA formed between two bodies under a load making contacts as shown (marked x in Fig. 6.1). Suppose now the normal load is increased and the line AA moves successively through rows of atoms 1, 2, 3 and 4 to establish a new equilibrium contact. At each row, let the number of atomic couples being repelled still be n_o. Then if the total distance traversed is x, the number of times the top surface will separate from the bottom surface is $(n_o x)/e$.

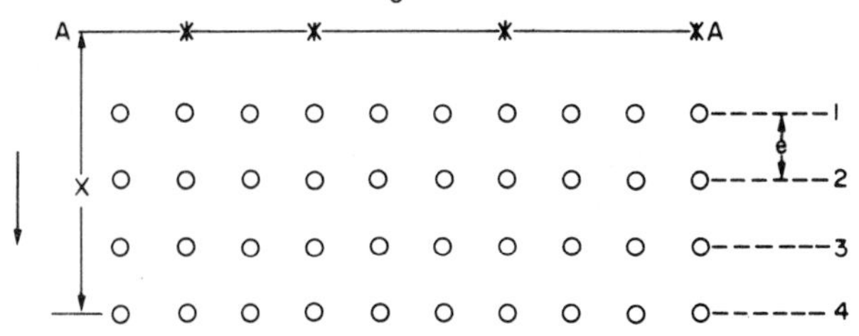

Fig. 6.1 An idealised model showing movement of one crystal through another. O represents an atom. The line AA is the interface formed making contact points marked X. The number of contacts is n_o and the same number, n_o, of atomic repulsion is assumed to take place as the line AA moves successively in the direction of the arrow. That is, while crossing row 1 there are n_o number of repulsions, the same number again while crossing row 2 and so on.

The assumed constancy of n_o is of course erroneous but Tomlinson recognises that a probability factor α has to be incorporated to account for the fact that in a real situation the total number of surface separations, n, for a distance x will be somewhat different. Thus

$$n = \alpha \frac{n_o x}{e} \qquad (6.4)$$

where α never exceeds unity.

Multiplying both sides by the mean energy E per atomic repulsion,

$$nE = \alpha \frac{n_o x}{e} E \qquad (6.5)$$

Comparing equations 6.3 and 6.5,

$$\mu P x = \alpha \frac{n_o x}{e} E$$

or

$$\mu = \alpha \frac{n_o}{eP} E$$

and substituting for P from equation 6.2

$$\mu = \frac{\alpha E}{e p_o} \qquad (6.6)$$

When an atom is displaced from its natural equilibrium state, it tends to return to its original position. However, during a separation it may enter the field of attraction of some other neighbouring atom before returning to its original position of equilibrium. Thus the atom in flight will undergo an attractive pull as dictated by the distance of separation ℓ at that instant and not much error will be introduced if this is regarded as equivalent to the interatomic force of cohesion F_o between two atoms in the crystal. The mean energy E for an atomic contact can, therefore, be expressed as $E = F_o\ell$, so that, from equation 6.6

$$\mu = \alpha \frac{F_o \ell}{e p_o} \tag{6.7}$$

The terms F and p_o are associated with the elastic constants of the material and equation 6.7 shows that the coefficient of friction is independent of the applied load which is Amontons' second law of friction.

6.2 *Wear*

The degree of proximity of two surfaces, that is their compliance, depends on statistical chance as the surfaces separate in the horizontal plane during sliding and try to mate due to the attractive force between their atoms. When sufficiently close, the atoms will be repelled and its natural trend is to return to its original position. However, it is a plausible hypothesis that an atom may be dislodged and move far enough to come within the field of another atom in the opposite surface where it finds a new equilibrium position. That is, atoms from one body can be plucked by others in the opposing surface. According to Tomlinson[1], this is the mechanism of wear.

From previous discussion, the energy dissipated by an atomic couple if $F_o\ell$. If ρ is the density of the metal which is wearing, that is whose atoms are being plucked, the mass of an atom is $m = \rho e^3$ (Fig. 6.1). If the total energy dissipated is E_t and the number of atomic junctions is n,

$$n = \frac{E_t}{F_o \ell}$$

The total mass of all the participating atoms which are being removed from the surface is

$$M = nm$$

or

$$M = \frac{E_t \rho e^3}{F_o \ell} \tag{6.8}$$

It should be noted that, although each junction has two atoms, only one atom is being plucked.

Now substituting for $F_o\ell$ from equation 6.7, equation 6.8 becomes

$$M = \alpha \frac{E_t \rho e^2}{\mu p_o} \tag{6.9}$$

It is known that each metal possesses its characteristic flow stress which is the limiting force the space lattice can withstand. The flow stress σ_y can be expressed in terms of the maximum repelling force p_{max} as

$$\sigma_y = \frac{p_{max}}{e^2}$$

For elastic contact, which is the model assumed here,

$$p_o = \frac{1}{2} p_{max}, \text{ so that } \sigma_y = \frac{2p_o}{e^2}$$

Substituting for e^2/p_o from this in equation 6.9,

$$M = \frac{2\alpha E_t \rho}{\mu \sigma_y} \tag{6.10}$$

Tomlinson's equation shows that the total mass of the metal removed is inversely proportional to the flow pressure.

REFERENCE

1. Tomlinson G A, *Phil Mag 7*, (1929), *7*, (46) 905.

CHAPTER 7

RUNNING-IN WEAR

Wear processes in metals have been classified into many types depending on the mechanism responsible for removal of material from surfaces. Thus corrosive wear implies the formation of loosely held fragments and their subsequent detachment when two bodies are in relative motion. Other forms of wear have been discussed viz., erosive and cavitation wear, fretting and fatigue wear. The definition and the nature of fretting are given in chapter 21 and cavitation, erosion and fatigue wear are briefly outlined in chapter 27. Whatever the nature of the geometry of movement or the environment, it would appear that most wear processes involve adhesion of asperities and the subsequent shearing of the junctions or a direct process of abrasion of a soft surface by a hard material. Both adhesive and abrasive wear have been discussed by many authors in the published literature.

7.1 *Wear Curve*

If volume or weight loss is plotted continuously with sliding distance a characteristic curve is obtained as shown in Fig. 7.1. The point 0 is when a machine started, that is when the sliding distance is zero. The weight loss is initially curvilinear and the rate of weight or volume loss per unit sliding distance decreases until at A where it joins smoothly with the straight line AB. The amount of volume loss in the regime given by OA is the running-in wear and AB is the steady state. The slope of the linear steady state regime is used to express the wear rate of a material per unit sliding distance and, at a given load and speed, it is constant for a material depending on the nature of the other surface. The pattern of the curve as in Fig. 7.1 has been shown to be the same in actual machine parts and there is due emphasis on the importance of running-in before a sliding component takes up its full load during its life of operation.

7.2 *Mechanism*

Surfaces under counterformal contact, for example a hemispherical pin on a bush, create a situation where the load is concentrated on a parallel narrow band of highly stressed metal.[1] The soft member of the couple will flow plastically and the area of contact will grow into an equilibrium size commensurate with the applied load. If both metals are hard the high spots will also be removed by some mechanism and again the surface undulations will take up a flat appearance as the area of true contact increases. That is, the surfaces have reached an equilibrium state and they have run in[2]. It is not certain if a run-in condition can be completely achieved once an equilibrium area of contact has been established. It is known[3] that gross surface flow occurs when sliding commences even at a moderate load causing an increase in the hardness of the interface. A work hardened zone also forms below the surface and there is evidence of formation of strongly adherent oxides which protect machine parts from gross distress in service. For example[4], electron diffraction shows that the run-in surfaces of cast iron piston rings and cylinders possess an oxide layer and there is the presence of graphite flakes which are oriented with their cleavage planes parallel to the surface. The mechanisms of running-in wear is probably governed by a large number of interrelated factors and should be a field for further research.

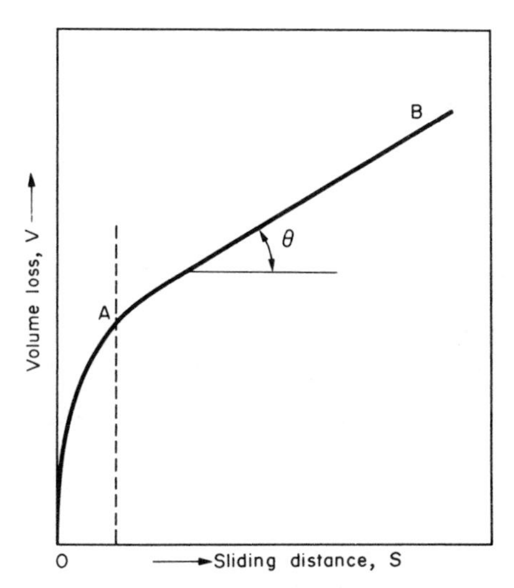

Fig. 7.1 The pattern of a typical wear curve OA is curvilinear and is the running-in wear. AB is linear and is the steady state wear. tan θ measures the rate of wear in loss of volume per unit sliding distance.

7.3 *Law of Running-In Wear*
A mathematical law[5] of running-in wear, also named as unsteady-state or transient wear, has been deduced along the following lines. At the instant of wear, the rate of volume removed per unit sliding distance must be a function of the volume of metal available at the junctions. That is, in general terms

$$\frac{dV}{dS} = -nV \tag{7.1}$$

where the V term denotes volume, S the sliding distance and n is a constant which depends, possibly, on the applied load. The negative sign describes a situation where the original volume at the junctions diminishes with sliding distance.

Integrating equation 7.1,

$$\ln V = -nS + C$$

where C is the constant of integration.

or

$$V = \exp(-nS + C) \tag{7.2}$$

If V_o is the volume generated at the junctions at zero sliding distance, putting S = o in equation 7.2,

$$V_o = \exp(C)$$

Therefore, from equation 7.2, the volume remaining at a sliding distance S is

$$V = V_o \exp(-nS)$$

Thus volume removed V_r at a sliding distance S is

$$V_r = V_o - V = V_o - V_o \exp(-nS)$$

or

$$V_r = V_o \left[1 - \exp(-nS) \right] \tag{7.3}$$

Now V_o is the volume generated by a load and should be, to a first approximation, equal to the sum total of the true contact areas, A_t, times the mean value of the change in compliance $\overline{\delta}$. The latter could be taken as the mean distance through which the centre line averages of the two bodies have come close together during compression by the normal load. Thus,

$$V_o = A_t \overline{\delta} \tag{7.4}$$

Combining equations 7.3 and 7.4, the volume removed during the running-in wear in terms of the sliding distance S is

$$V_r = A_t \overline{\delta} \left[1 - \exp(-nS) \right] \tag{7.5}$$

The term A_t is related to the load and flow pressure σ_y of the material, so that

$$V_r = \frac{W}{\sigma_y} \overline{\delta} \left[1 - \exp(-nS) \right] \tag{7.6}$$

$\overline{\delta}$ could, probably, be estimated from surface traces. However, n is unknown and is likely to be a function of load, the flow stress of the material which is wearing and the surface roughness of one or both members of the couple.

REFERENCES

1. Barwell F T, *Proc Inst Mech Engrs (London)*, (1967-68), *182*, (3K), 1.
2. Burwell J T, *Wear*, (1957-58), *1*, 119.
3. Bowden F P and Hughes T P, *Proc Roy Soc A*, (1937), *160*, 575.
4. Finch G L, *Proc Roy Soc A*, (1950), *63*, 785.
5. Queener C A, Smith T C and Mitchell W L, *Wear*, (1965), *8*, 391.

CHAPTER 8

ADHESIVE WEAR

It is clear that two interacting surfaces make contact only at a few isolated points resulting in high stresses in these areas. This gives rise to plastic flow at the interface and chapter 4 showed that the resistance to sliding is equivalent to the sum of the shearing forces necessary to break up all these junctions, assuming the absence of a ploughing term. Whereas friction is accounted for this way, a general law of wear which could be used by design engineers is not yet complete, possibly because wear is a complex process depending on, inter alia, normal load, crystal structure and mechanical properties of materials. Nevertheless, laws of wear have been formulated and much attention has been paid to adhesive wear which can be defined as being that process which gives rise to loss of metal between interacting surfaces as a result of adhesion of asperities. Wear by this mechanism is severe and can happen in the kinematic chain of a machine whenever there is an absence of effective lubrication at the interface of two bodies in contact.

8.1 *Rate of Wear*

It was shown in chapter 6 that Tomlinson enunciated a model where surface atoms were plucked and removed by the combined action of the cohesive and repulsive forces existing in metal crystals. It was deduced that wear was inversely proportional to the flow pressure σ_y of the metal.

Holm[1], using the same basic consideration, proposed that as sliding commences, atom to atom contact removes surface atoms at favourable encounters so that the loss of volume V for a sliding distance S is

$$V = ZA_tS \tag{8.1}$$

where A_t is the true area of contact and Z is the number of atoms removed per encounter. The implication of a true area of contact is the same as in the ideas of the laws of friction and

$$A_t = \frac{W}{\sigma_y} \tag{8.2}$$

where W is the applied load and σ_y is the flow pressure of the softer metal. Substituting for A_t from equation 8.2 in equation 8.1

$$\frac{V}{S} = Z\ \frac{W}{\sigma_y} \tag{8.3}$$

The term V/S is the volume rate of wear per unit sliding distance and is seen to be inversely proportional to the flow stress, being similar to the law of Tomlinson derived from basic consideration.

Equation 8.3 states that the total volume of material removed due to sliding is proportional (a) to the applied normal load; (b) to the sliding distance and (c) inversely to the flow pressure of the material.

Both the laws of Tomlinson and Holm show that the volume of metal removed during surface interaction is independent of the apparent contact area.

8.2 *Junction Interaction*

The idea of metal removal by plucking of atoms has been disputed by many on the ground that wear debris are invariably aggregates of smaller metallic particles. This suggests that, at an encounter, a fraction of the surface at the interface is removed which is many times larger than the size of a single atom.

Burwell and Strang[2] have examined Holm's equation (equation 8.3) by running conical brass and steel pins on steel discs. Upon plotting the volume of metal removed against sliding distance, a running-in period is observed followed by a steady state wear (chapter 7). Taking the slope of the latter as the rate of wear, it is observed that this is proportional to the apparent normal contact pressure up to a value which equals the yield stress in tension of the pin material. Beyond this, equation 8.3 is not obeyed and the rate of wear increases many fold. In this regime of severe wear, according to the authors[2], a wear debris produced at the leading edge causes scoring and metal removal from the pin further downstream. As sliding continues, fresh fragments form which cause further attrition and produce more debris so that an avalanche of wear particles is produced this way.

A modified law of wear is formulated[2] by replacing Z in equation 8.3 with β_1, the probability of producing a wear fragment at an encounter, so that

$$V = \beta_1 \frac{W}{\sigma_y} S \qquad (8.4)$$

Dividing both sides by the apparent area of contact,

$$h = \beta_1 \frac{\sigma_a}{\sigma_y} S \qquad (8.5)$$

where h is the height loss due to wear and σ_a is the apparent contact pressure exerted by the pin on the disc. The authors[2] have plotted (Fig. 8.1) $h/\sigma_a S$ against σ_a and a large increase in the value of $h/\sigma_a S$ is obtained at a contact pressure equivalent to 1/3rd the hardness of the pin. That is, wear becomes catastrophic when the normal load is such that the flow pressure of the brass pin is reached. This is discussed again in chapter 23 in the light of experimental results obtained with a pin-bush machine.

8.3 *Law of Adhesive Wear*

In the model proposed by Archard[3], two nominally flat surfaces make contact at the high asperities which flow plastically because of the concentrated localised stresses there. At the lightest load, contact occurs at three points only and, as the external stress is increased, the original contact area becomes larger. As this happens, the compliance between the couple improves, that is the gap between the two surfaces diminishes resulting in further protuberances making contacts elsewhere.

The idea of single atoms being removed from the surfaces is discarded, that

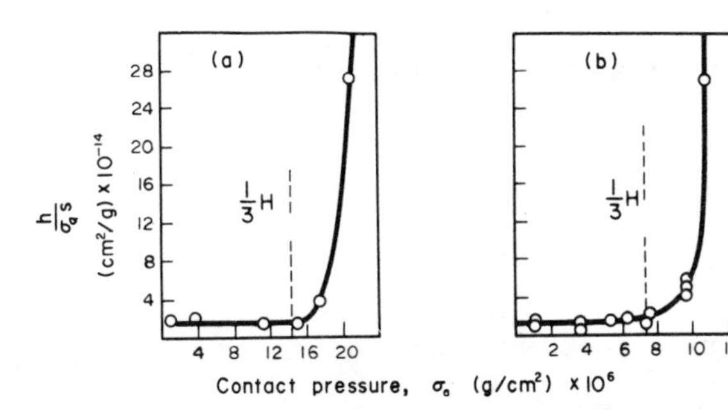

Fig. 8.1 Rate of wear as a function of apparent contact pressure (Burwell and Strang[2]), h = height loss of pin due to wear; s = sliding distance. Conical steel pins on steel disc with pin hardness in kg/mm^2, (a) 430 (b)223.

is, it is assumed that wear occurs by removing fragments of metal from the surface. Assuming a wear fragment to be hemispherical with an average diameter of 2r, the true area of contact A_t is

$A_t = n\pi r^2$, where n is the number of junctions.

Since $A_t = W/\sigma_y$,

$$n = \frac{A_t}{\pi r^2} = \frac{W}{\sigma_y \pi r^2} \tag{8.6}$$

For circular junctions of diameter 2r, the distance traversed for breaking one junction by a tangential pull can be regarded as 2r. In a situation like this, one body is moving a distance 2r relative to the other surface. Since the total number of asperities involved is n, the number of asperities per unit sliding distance is

$$n_u = \frac{n}{2r}$$

or, from equation 8.6

$$n_u = \frac{W}{2\sigma_y \pi r^3} \tag{8.7}$$

It does not follow that all the junctions which break will form a wear fragment. It is an important argument since it is known that a high friction will not necessarily mean a proportionately high wear. Thus whereas breaking of each junction will result in a resistance component to sliding, it does not follow that it will detach from the metal surface giving a wear debris. It may simply mean that as the junction breaks it remains attached to the metal surface as microscopically rough areas or, if plastic, may be just smeared over the metal surface.

Therefore, assume that there is a probability β that for n_u junctions per unit sliding distance, a number of wear fragments will form. Then for a larger distance S giving a total volume of wear debris V, the rate of wear per unit sliding distance can be represented as

$$\frac{V}{S} = \beta\, n_u \times \text{(volume of a hemispherical wear fragment)} \tag{8.8}$$

$$= \beta\, n_u \frac{2}{3} \pi r^3$$

And substituting for n_u from equation 8.7,

$$\frac{V}{S} = \beta \frac{W}{3\sigma_y}$$

or
$$V = \beta \frac{W}{3\sigma_y} S \tag{8.9}$$

The equation is similar to those proposed by Tomlinson and in particular by Holm and the volume of the wear debris is shown to be independent of the apparent area of contact. The size of the junctions does not enter into it but there is a shape factor, viz., 3 for a hemispherical asperity. The shape factor is different for a cube or a cylinder.

8.4 *Asperity Angle*

Real surfaces are not a regular array of hemispherical protuberances but have shapes probably between a cone and a hemisphere. The wear rate of metals, assuming conical model surfaces, has been deduced[4] as follows.

Assume an interface (Fig. 8.2) where the top surface is perfectly flat but the bottom surface is undulated with metal asperities of a conical nature distant o, h, 4h, 2h, 3h etc. from the flat surface. Assume also that the asperities are randomly distributed in space and each has a base angle θ. At the instant of no load the reference line is $z = o$ but if the tips of the asperities are squashed due to the contact stress, the top surface moves relative to $z = o$ in the z direction as indicated in Fig. 8.2.

Let n = total number of asperities, and

σ_y = flow pressure of the bottom material in Fig. 8.2.

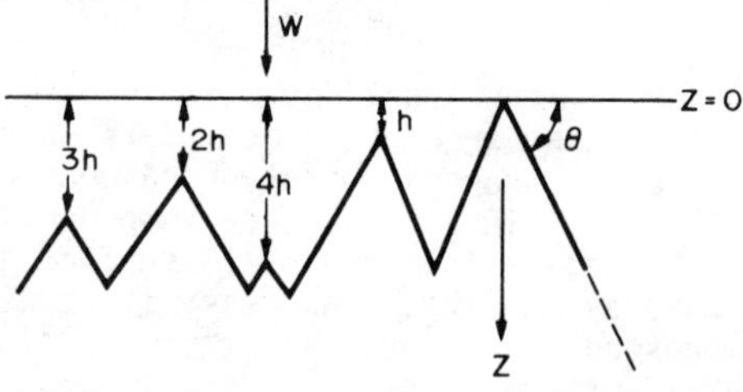

Fig. 8.2 A flat surface approaching another with conical asperities (after Yoshimoto and Tsukizoe[4]).

The simplified model shows that only the apex of one asperity touches the top surface so that the apices of all the n asperities are at distances $\left[r = 0, 1, 2, 3, 4, \ldots (n-1) \right] h$ respectively.

Depending on the load, an asperity will be flattened to a basal diameter 2r and the top surface is assumed to remove this deformed portion of the cone as sliding is completed through a distance 2r. This is shown schematically in Fig. 8.3.

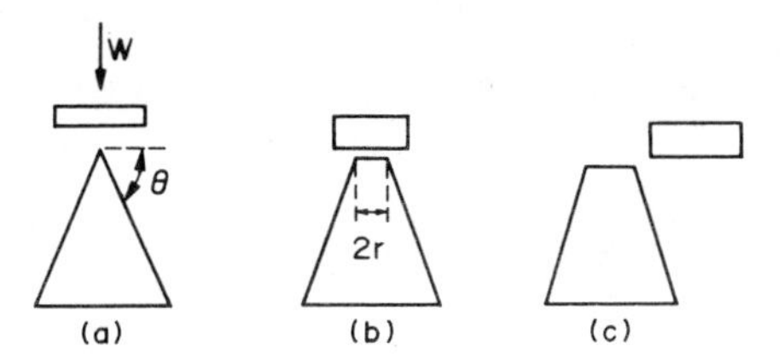

Fig. 8.3 Schematic representation of a conical asperity being deformed and worn by a flat surface. (a) flat surface approaching; (b) deformation of the conical asperity; (c) flat surface has slid past the area of contact producing wear debris.

It is thus assumed that the sliding surfaces first make a contact of diameter 2r and then move on to make a contact elsewhere of the same diameter since the load is constant and the process continues until all the n asperities have interacted. Now if h_o is the height of the portion of a conical asperity which has been removed, the volume ΔV of metal lost per asperity is

$$\Delta V = \frac{1}{3} \pi r^2 h_o \tag{8.10}$$

But $h_o = r \tan \theta$.

Substituting this in equation 8.10

$$\Delta V = \frac{1}{3} \pi r^2 \, r \tan \theta = \frac{1}{3} \pi r^3 \tan \theta$$

Therefore, the volume removed per unit sliding distance ΔV_o is

$$\Delta V_o = \left\{ \frac{1}{3} \pi r^3 \tan \theta \right\} / 2r$$

$$\Delta V_o = \frac{1}{6} \pi r^2 \tan \theta$$

Working on the premise that the top surface moves through a distance equal to the diameter 2r of the contact area relative to the bottom surface, the volume wear per unit sliding distance due to n asperities is

$$V = \sum_{r=o}^{n-1} \frac{1}{6} \pi r^2 \tan \theta$$

$$= \frac{\tan \theta}{6} \sum_{r=o}^{n-1} \pi r^2$$

Now $\sum_{r=o}^{n-1} \pi r^2 = A_t$, the true area of contact,and

$$A_t = \frac{W}{\sigma_y}$$

Therefore, $$V = \frac{1}{6} \frac{\tan \theta}{\sigma_y} W \qquad (8.11)$$

Equation 8.11 is similar to equation 8.9 in that the wear volume is directly proportional to the load W and inversely to the flow pressure of the metal that is undergoing wear.

The model is applicable to adhesive wear or what the authors[4] termed as mechanical wear.

8.4.1 *Hemispherical Asperity.* Using the same model, and assuming contact with a hemispherical asperity and again assuming that a part of this is removed, the volume removed per asperity is

$\frac{2}{3} \pi r^2$, where r is the diameter of the contact area.

Then, if the relative movement of the surfaces is 2r,

Volume of wear/unit distance is $\frac{1}{3} \pi r^2$

The total volume removed per n asperities is

$$V = \frac{1}{3} \sum_{r=o}^{n-1} \pi r^2$$

$$= \frac{1}{3} A_t = \frac{1}{3} \frac{W}{\sigma_y}$$

Therefore, if the total distance slid is S,

$$V = \frac{W}{3\sigma_y} S \qquad (8.12)$$

This is very similar to Archard's law but the authors[4] assume that a wear debris is produced at each encounter which is not so in practice.

8.5 *Fatigue Mechanism*
An important aspect of tribological surfaces is that they soon work harden to a depth and it is probable that the interaction is then largely elastic. Since the maximum Hertzian shear stress is below the surface, it is probable that a fatigue crack would nucleate there and propagate to the surface to produce a wear debris. This fatigue mechanism could be the main influence in causing wear during the steady state and Halling[5] has proposed a mathematical model incorporating the concept of fatigue failure in metals.

The author considers a rigid rough surface pressed against another consisting of an array of spherical asperities of a material defined by equation 3.16 (Fig. 8.4) The asperities are assumed to be spherical each with a radius R. The true area of contact and the normal load are expressed in the same manner as in section 3.6.

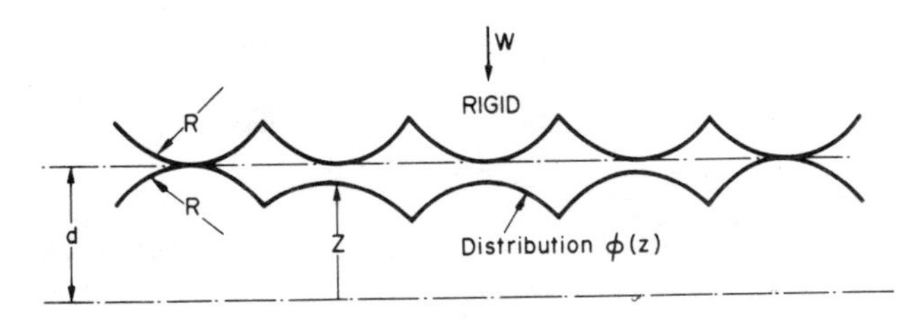

Fig. 8.4 The normal contact of rough surfaces (Halling[5]).

8.5.1 *Contact of Sliding Surfaces.* Figure 8.5a shows the situation when two asperities have just made contact and Fig. 8.5b shows the situation when the bottom asperity has been deformed by a maximum amount. It is assumed that the deformation and strain are mostly dependent on the normal load, that is the effect of the tangential force is neglected. The author[5] uses a failure criterion which requires a knowledge of the maximum plastic strain induced in the deforming material during each asperity contact cycle. From equation 3.21, since $a = \lambda^{\frac{1}{2}}R^{\frac{1}{2}}\delta^{\frac{1}{2}}$, the strain $\bar{\varepsilon}$ is maximum when the compliance δ is a maximum. From Fig. 8.5,

$$\delta_{max} = 2R - b$$

Therefore
$$\bar{\varepsilon}_{max} = \left(\frac{K}{\pi BC}\right)^{\frac{1}{P}} \lambda^{\frac{1}{2}} \left(2 - \frac{b}{R}\right)^{\frac{1}{2}} \qquad (8.13)$$

For an asperity of initial height z (Fig. 8.5)

$$b = d + 2R - z$$

Therefore

$$\bar{\varepsilon}_{max} = \left\{\frac{K}{\pi BC}\right\}^{\frac{1}{P}} \left\{\frac{\lambda}{R}\right\}^{\frac{1}{2}} (z-d)^{\frac{1}{2}} \tag{8.14}$$

If $\bar{\varepsilon}_1$ is the maximum plastic strain at failure after one cycle of loading and N_+ is the total number of contacts to produce failure,

$$\left[\frac{\bar{\varepsilon}_1}{\bar{\varepsilon}_{max}}\right]^m = N_+ \tag{8.15}$$

The index m is usually taken as 2 for metals.

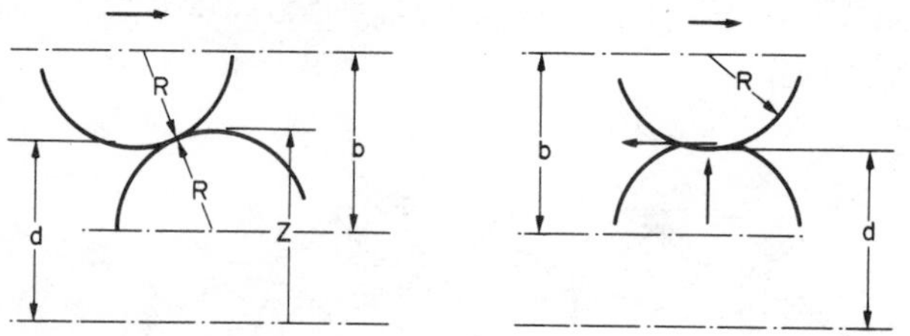

Fig. 8.5 The asperity contact during sliding. (a) Initial contact; (b) Maximum deformation (Halling[5]).

8.5.2 *The Wear Law.* Consider now the surfaces to undergo relative sliding at a constant separation distance d (Fig. 8.6). Consider a small width w as shown in Fig. 8.6 and let the distance traversed be S. Let n_u be the total number of asperities in one row on the upper surface and n_s the number of asperities on the lower surface along the direction of motion over the distance S. If $\phi(z)$ is the probability value for asperities of height z on the lower surface, the number of contacts of an asperity over a distance S is

$$n_u n_s \phi(z)$$

suppose that a wear particle of volume V is produced after N_+ contacts. Assume also that the size of the detached particle is related to the maximum strain, $\bar{\varepsilon}_{max}$, during the asperity contact as

$$V = \gamma \bar{\varepsilon}_{max}^{\,t} \tag{8.16}$$

where γ is a constant.

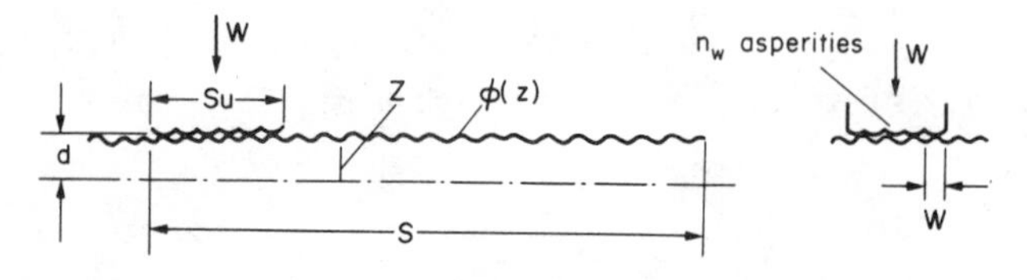

Fig. 8.6 The model used for the derivation of a wear law (Halling[5]).

If n_w is the number of rows of asperities along the width of the upper surface, the total volume of wear product is

$$V = N_w n_u n_s \gamma \int_d^\infty \frac{\bar{\varepsilon}_{max}^{\ t} \ \phi(z)dz}{N_+} \tag{8.17}$$

Now $$n_u = ns_u \tag{8.18}$$

$$n_s = nS \tag{8.19}$$

where n is the number of asperities per unit length for both surfaces; s_u and S are the lengths involved for the upper and lower surfaces respectively.

Substituting for N_+ from equation 8.15 in equation 8.17

$$V = n_w n_u n_s \gamma \int_d^\infty \frac{\left(\bar{\varepsilon}_{max}\right)^{t + m}}{\left(\bar{\varepsilon}_1\right)^m} \phi(z)dz$$

Substituting in the above relationship for $\bar{\varepsilon}_{max}$ from 8.14,

$$V = \frac{n_w n_u n_s \gamma}{\bar{\varepsilon}_1^{\ m}} \int_d^\infty \left(\frac{K}{\pi BC}\right)^{\frac{m + t}{p}} \left(\frac{\lambda}{R}\right)^{\frac{m + t}{2}} \left[z\text{-}d\right]^{\frac{m + t}{2}} \phi(z)dz$$

And substituting for n_u and n_s from equations 8.18 and 8.19,

$$\frac{V}{S} = \frac{n_w n^2 s_u \gamma}{\bar{\varepsilon}_1^{\ m}} \left(\frac{K}{\pi BC}\right)^{\frac{m + t}{p}} \left(\frac{\lambda}{R}\right)^{\frac{m + t}{2}} \int_d^\infty (z\text{-}d)^{\frac{m + t}{2}} \phi(z)dz \tag{8.20}$$

Following the principles given in chapter 3, considering the upper 25% of a real surface to have an exponential height distribution, the volume loss per unit sliding distance, that is the wear rate V/S can be written as

$$\frac{V}{S} = \left[\frac{n}{\pi\bar{\varepsilon}_1^{\,m}} \; \frac{(\sigma\lambda)^{\frac{t - p + m - 2}{2}}}{(R)^{\frac{2 + m - p + t}{2}}} \left(\frac{K}{\pi BC}\right)^{\frac{m + t}{p} - 1} \frac{\left(\frac{m + t}{2}\right)!}{\left(1 + \frac{p}{2}\right)!} \right] \frac{W}{CB} \qquad (8.21)$$

or

$$\frac{V}{S} = C_1 \frac{W}{CB} \qquad (8.22)$$

where C_1 is constant which equals the terms in [ ] in equation 8.21. The author[5] argues that the wear rate in equation 8.22 can be expressed in terms of H_i, the initial hardness of the material, so that

$$\frac{V}{S} = C_2 \frac{W}{H_i} \qquad (8.23)$$

where C_2 is a constant.

Considering equations 8.21 and 8.23 the amount of wear is proportional directly to the sliding distance and inversely to the prior hardness of the metal. Examining C_1, for the proposed wear law to have any meaning, $p \neq o$, that is the system is non-perfectly-plastic. Both t and p are less than unity and m = 2, so that σ, the RMS value of surface texture, is not significant. The most significant surface parameters are the asperity density and the size of the summits as defined by the asperity radius.

REFERENCES

1. Holm R, *'Electrical Contacts'*, H Gerbers, Stockholm, (1946).
2. Burwell J T and Strang C D, *Jl Appl Physics*, (1952), *23*, 18.
3. Archard J F, *Jl Appl Physics*, (1953), *24*, 981.
4. Yoshimoto G and Tsukizoe T, *Wear*, (1958), *1*, (6), 472.
5. Halling J, *Wear*, (1975) *34*, 239.

CHAPTER 9

OXIDATIONAL HYPOTHESIS OF WEAR

Laws of adhesive wear (chapter 8) assume first and foremost that surface interaction occurs in a perfectly clean environment. That is, there is a complete absence of any contaminants. Chemically clean conditions can only prevail in a perfect vacuum and, in room atmosphere, surfaces are inevitably covered with layers of sorbed gases. Above all, it is an ubiquitous feature of most metals that a layer of oxide rapidly grows on them if they are exposed to an oxygen bearing environment. Presence of oxides means that the integrity of the junctions is disrupted which, by itself, should give diminished wear. This is so in practice and reference to the role of oxides from the viewpoint of wear will again be made in subsequent chapters.

9.1 *Oxidational Hypothesis*

Equation 8.9 has been accepted as the law of adhesive wear but with the assumption that the interface is clean. Since this is not so in practice, the value of the probability β of forming a wear fragment should be modified. Quinn[1] has derived a relationship for β in terms of some basic parameters for wear of steels as follows.

Assuming mechanical wear, $1/\beta$ encounters are necessary for a critical oxide film thickness ξ to build up before it is dislodged to form a wear particle. It would appear that the hypothesis assumes that, at each encounter, frictional heat promotes the formation of oxide until, after a few encounters, the oxide film is thick enough to be detached from the body of the metal. The film is, possibly, removed by simple ploughing by a hard asperity or its detachment from the substrate may be explained by the work of adhesion concept (chapter 5).

9.1.1 *The* β *Term.*

If Δt is the time of contact for a single encounter, the total time t to build up a critical thickness is

$$t = \Delta t/\beta \tag{9.1}$$

Also, if d is the sliding distance for a single encounter and v the velocity,

$$\Delta t = d/v$$

substituting this value of Δt in equation 9.1,

$$t = \frac{d}{v\beta} \tag{9.2}$$

The law of oxidation states that the growth of a mass of oxide per unit area, Δm, is given by

$$\Delta m^2 = Kt, \text{ where K is a constant}$$

Assuming $\Delta m = \xi\rho$, where ρ is the density of the oxide film,

$$\xi^2 = Kt/\rho^2$$

i.e.

$$t = \frac{\xi^2\rho^2}{K} \tag{9.3}$$

comparing equations 9.2 with 9.3,

$$\frac{\xi^2\rho^2}{K} = \frac{d}{v\beta}$$

$$\beta = \frac{Kd}{v\xi^2\rho^2} \tag{9.4}$$

It is known that

$$K = A_0 \exp(-E/RT) \tag{9.5}$$

where
A_0 = Arrhenius constant
E = Activation energy
R = Universal gas constant
T = Absolute temperature of the sliding interface

Substituting the value of K from equation 9.5 in equation 9.4,

$$\beta = A_0 \exp(-E/RT)d/v\xi^2\rho^2 \tag{9.6}$$

The full equation of wear can be written as

$$\frac{V}{S} = \left[A_0 \exp(-E/RT)d/v\xi^2\rho^2\right] \frac{W}{\sigma_y} \tag{9.7}$$

where the symbols have the same meaning as in chapter 8.

9.2 *Comment on Equation 9.7.*
Since the volume rate of wear per unit sliding distance is directly governed by β, a high value of this means an increase in the rate of wear and vice versa. The term d in equation 9.6 can be regarded, for all practical purposes, as the diameter of the contact area if it is circular and is probably decided by the normal stress. It is known that increased sliding velocity results in a rise in surface temperature which facilitates oxide formation. The term ξ^2 is difficult to explain since equation 9.7 suggests that the thicker the oxide film the lower the rate of wear. However, the assumption in 9.1 is that the metal loses its volume as the oxide film thickens to a critical value and then detaches. It is more reasonable to infer that if the film is thin it will remain attached to the surface and volume

loss from the surface will be restricted. However, the protective effect of oxide film with regard to surface damage and wear is well known.

REFERENCE

1. Quinn T F J, *'The Application of Modern Physical Techniques to Tribology'*, Newnes-Butterworths, (1971).

CHAPTER 10

SURFACE CONTAMINANTS

The more clean the interface, the greater is the degree of adhesion between two bodies so that seizure between them due to strong interfacial adhesion is experienced in high vacuum. Bryant[1] shows that the energy required to cause cleavage of mica layers in a vacuum of 10^{-13} torr is some 30 times greater than that in air. Contamination of surfaces occurs easily from the surrounding atmosphere and, indeed, this is beneficial having regard to the life of a component from the point of view of wear or its failure by seizure. Since a prerequisite of adhesion of surfaces is intimate metal to metal contact, a contaminant, in the same manner as an oxide film, will dilute the junctions and thus reduce the propensity to adhesive wear.

10.1 *Fractional Film Defect*

Rowe[2] suggests that Holm's equation for adhesive wear (chapter 8) should be modified by incorporating a factor α to take account of the presence of surface contaminants. By his definition, $\alpha = A_t/A$ where A_t and A are the true and apparent areas of contact respectively and the ratio has a value between zero and unity. Thus the modified equation for the volume of material removed V for a sliding distance S is

$$V = \alpha\beta A_t S \tag{10.1}$$

It is worth observing at this stage that if A_t is zero, there is no wear, but for α to be unity, the apparent area must equal the total area of the junctions. The β term is still a valid proposition because it does not follow that, if true contact is established throughout the whole apparent area, there will be wear each time there is relative movement between the two surfaces.

10.1.1 *The Term* α. The basic equation of adhesive wear (equation 8.9) has already stated that β is the probability of forming a debris at an encounter. The apparent area does not equal the true contact area, the magnitude of the latter depending on the applied load W as defined in chapter 8. The term α has been named by Rowe[2] as the fractional surface film defect and must mean that the true area of contact is a factor α times the ratio W/σ_y.

10.1.2 *Modified Wear Equation.* For a two-dimensional stress system, the material will undergo plastic flow when

$$\alpha^2 + 3\tau^2 = \sigma_y^2 \tag{10.2}$$

where σ is the applied normal stress, τ the shear stress at the junction and σ_y is the flow pressure.

The frictional resistance F of an interface is related to the true area of contact A_t as

$$F = \tau A_t \tag{10.3}$$

since $\mu = F/W = \tau A_t/W$, $\tau = \mu\sigma$

substituting this value of τ in equation 10.2,

$$\sigma^2 + 3(\mu^2\sigma^2) = \sigma_y^2$$

or $$\sigma = \frac{W}{A_t} = \frac{\sigma_y}{(1 + 3\mu^2)^{\frac{1}{2}}}$$

and $$A_t = \frac{W(1 + 3\mu^2)^{\frac{1}{2}}}{\sigma_y}$$

substituting this value of A_t in equation 10.1,

$$V = \alpha\beta(1 + 3\mu^2)^{\frac{1}{2}}\left(\frac{W}{\sigma_y}\right)S \tag{10.4}$$

Since the β term can only be obtained empirically, at a first glance it would appear pointless to pursue equation 10.4. However, it is probable that the term β for any couple should be obtained by conducting experiments in vacuum which gives the wear potential for complete metal to metal interaction. This should be compared with experiments in air or any desired atmosphere to evaluate α which is a factor describing the degree of contamination of the sliding interface. An aspect of equation 10.4 is that it suggests a dependence of wear on the coefficient of friction. One of the reasons for a lack of correlation between the rate of wear and the coefficient of friction is, possibly, the effect of the factor α.

10.2 *Heat of Adsorption Theory*

Wear of metals has been related to the heat of adsorption of a molecule of the contaminant notably by Kingsbury[3,4] and also by Rowe[2]. Thus consider an asperity A of one body moving at a velocity v in a sliding situation with a surface B (Fig. 10.1). An array of lubricant molecules has settled on surface B, shown as open circles in Fig. 10.1. Suppose the lubricant molecules are spread over the surface B such that the spacing z between them is the same as the lattice spacing of the material B. An adsorbed molecule does not remain on the same site for long but is desorbed, the vacancy thus left being filled by a neighbouring molecule. The average time t_r spent by a lubricant molecule on B is given by

$$t_r = t_o \exp(E/RT) \tag{10.5}$$

where t_o is the fundamental period of vibration of the lubricant molecule on B,

E is the heat of physical adsorption of a molecule normal to the surface B,

R is the universal gas constant,

T is the absolute temperature.

The fractional film defect α is defined as

$$\alpha = (N_s - N)/N_s \tag{10.6}$$

where N_s is the total number of sites on the surface B and N is the number of sites occupied under the asperity A.

Suppose that there is no interaction between a molecule of lubricant and the asperity A until the latter approaches the former to one lattice spacing z. The time required, t_z, for the asperity to move this distance is

$$t_z = z/v \tag{10.7}$$

If $t_z >> t_r$, the adsorbed molecule has ample time to desorb and hence $\alpha \to 1$, that is N is zero in equation 10.6. Conversely, if $t_z << t_r$, $\alpha \to 0$. These conditions are well satisfied by assuming a relationship of the following form:

$$(1 - \alpha) = \exp(-t_z/t_r) \tag{10.8}$$

The term α in equation 10.6 is in effect the fraction of the surface that is unlubricated. It follows that the fraction which is lubricated is given by $(1 - \alpha)$.

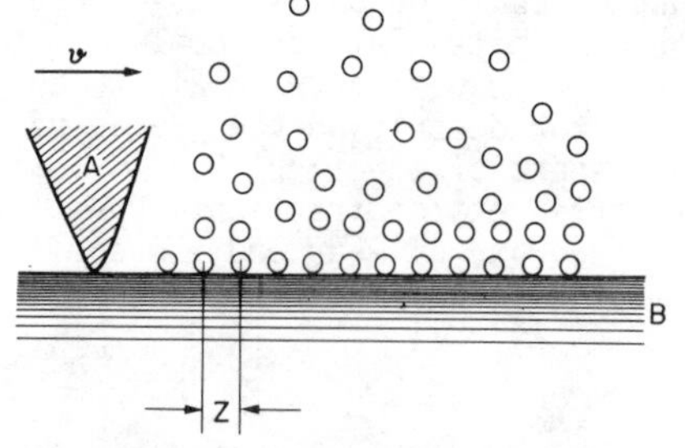

Fig. 10.1 A schematic model of an asperity A moving with a velocity v over a surface B covered with lubricant molecules having z as the distance of separation between them (after Kingsbury[2]).

10.2.1 *Friction.* It has been argued[5] that the net coefficient of friction under the condition of partial lubrication is the sum of the coefficient of friction due to metal contact and that due to the frictional force necessary to shear the lubricant itself. That is,

$$\mu = \alpha\mu_c + (1 - \alpha)\mu_\ell \tag{10.9}$$

where μ = the measured coefficient of friction

μ_c = the coefficient of friction between thoroughly cleaned surfaces

μ_ℓ = the coefficient of friction when the interface is completely partitioned by a layer of lubricant.

Substituting for α and $(1 - \alpha)$ from equation 10.8,

$$\mu = \left\{1 - \exp(-t_z/t_r)\right\} \mu_c + \left\{\exp(-t_z/t_r)\right\}\mu_\ell$$

Taking the values of t_r and t_z from equations 10.5 and 10.7 respectively,

$$\mu = \left\{1 - \exp\left[- z/vt_o \exp(E/RT)\right] \right\} \mu_c + \left\{\exp \left[-z/vt_o\exp(E/RT)\right] \right\}\mu_\ell$$

That is,

$$\mu = \left\{1 - \exp \left[- \frac{z}{vt_o} \exp(-E/RT)\right] \right\}\mu_c + \left\{\exp \left[- \frac{z}{vt_o} \exp (-E/RT)\right] \right\} \mu_\ell$$

or

$$\mu = \mu_c - \mu_c \left\{ \exp \left[- \frac{z}{vt_o} \exp(-E/RT)\right]\right\} + \mu_\ell\left\{\exp \left[- \frac{z}{vt_o} \exp(-E/RT)\right] \right\}$$

or

$$\mu_c - \mu = \left\{ \exp \left[- \frac{z}{vt_o} \exp (-E/RT)\right] \right\}(\mu_c - \mu_\ell)$$

or

$$\frac{\mu_c - \mu}{\mu_c - \mu_\ell} = \exp \left[- \frac{z}{vt_o} \exp (-E/RT)\right] \qquad (10.10)$$

Taking logarithms,

$$\ln (vt_o/z) = \ln \left\{\left[\ln\left(\frac{\mu_c - \mu_\ell}{\mu_c - \mu}\right)\right]^{-1} \right\} - \frac{E}{RT} \qquad (10.11)$$

Putting $Y = \ln(vt_o/z)$ and

$$X_f = \ln\left\{ \left[\ln\left(\frac{\mu_c - \mu_\ell}{\mu_c - \mu}\right) \right]^{-1} \right\}$$

$$Y = X_f - \frac{E}{RT} \qquad (10.12)$$

10.2.2 *Wear*. The concept of fractional film defect in terms of friction can also be applied to the situation of wear. Thus assuming adhesive wear, use a new notation λ for the fractional film defect while considering wear only. Let

q = observed rate of wear per unit sliding distance

q_c = wear rate when the metal surfaces are thoroughly clean

q_ℓ = rate of wear when the interface is fully lubricated.

Using the analogous situation to equation 10.9,

$$q = \lambda q_c + (1 - \lambda) q_\ell \tag{10.13}$$

However, if the surfaces are separated by a lubricant, there is no adhesion between the metal asperities and, therefore, there is no adhesive wear. That is, q_ℓ is zero in equation 10.13, so that

$$q = \lambda q_c \tag{10.14}$$

The film defect α is considered an area term whereas λ is a volume term[7], so that it is reasonable to state that

$$\alpha = (\lambda)^{\frac{2}{3}} \tag{10.15}$$

From equation 10.14,

$$\frac{q}{q_c} = \lambda$$

therefore, from equation 10.15,

$$\frac{q}{q_c} = \alpha^{\frac{3}{2}}$$

or substituting for α from equation 10.9,

$$\frac{q}{q_c} = \left(\frac{\mu - \mu_\ell}{\mu_c - \mu_\ell}\right)^{\frac{3}{2}}$$

that is

$$\frac{\mu - \mu_\ell}{\mu_c - \mu_\ell} = \left(\frac{q}{q_c}\right)^{\frac{2}{3}}$$

or

$$1 - \left(\frac{\mu - \mu_\ell}{\mu_c - \mu_\ell}\right) = 1 - \left(\frac{q}{q_c}\right)^{\frac{2}{3}}$$

or

$$\frac{\mu_c - \mu}{\mu_c - \mu_\ell} = 1 - \left(\frac{q}{q_c}\right)^{\frac{2}{3}}$$

substituting this in equation 10.11,

$$\ln (vt_0/z) = \ln \left\{ \left[\ln \left\{ \frac{1}{1 - \left(\frac{q}{q_c}\right)^{\frac{2}{3}}} \right\} \right]^{-1} \right\} - E/RT \qquad (10.16)$$

or making similar substitutions as in 10.12,

$$Y = X_w - E/RT \qquad (10.17)$$

10.3 *Importance of E.*
The heat of adsorption theory suggests that a very important parameter in a sliding situation is the fractional film defect α. Both friction and wear are low when $\alpha \to o$, that is when all the possible sites on a metal surface are occupied by lubricant molecules. As α increases, friction and, in particular, wear increase since metallic contact is facilitated. A very important property of a lubricant is its heat of physical adsorption, E. A high value of E means that α will be low and vice versa since a lubricant has a greater chance of staying on a surface if it has a high value of the heat of physical adsorption.

10.4 *A Simplified Law.*
Equation 10.16 is an important contribution which incorporates fundamental parameters of a lubricant. Unfortunately, it is unwieldy and alittle tedious to compute manually. Rowe[2] suggests that α usually has a low value of the order of 0.01 or less so that a good approximation is

$$\ln (1 - \alpha) = (-\alpha) \qquad (10.18)$$

It is emphasised, however, that caution is necessary since, at very low sliding speeds, e.g., 0.001 cm/s or at very high interface temperatures, α could be very high in which case the assumption given by equation 10.18 is invalid.

Now using equations, 10.5, 10.7 and 10.8,

$$\alpha = 1 - \exp \left[\frac{z}{vt_0} \exp \quad (-E/RT) \right]$$

Taking logs,

$$\alpha = \frac{z}{vt_0} \exp (-E/RT)$$

It is more realistic to replace z by d, the diameter of the area of contact, assuming it to be circular. Thus

$$\alpha = \frac{d}{vt_0} \exp(-E/RT) \qquad (10.19)$$

Substituting $\gamma = (1 + 3\mu^2)^{\frac{1}{2}}$ in equation 10.4,

$$V = \alpha\beta\gamma\left(\frac{W}{\sigma_y}\right)S \qquad (10.20)$$

Substituting in equation 10.20, the value of α from equation 10.19

$$\frac{V}{S} = \left(\frac{\beta\gamma d}{vt_o}\right)\left(\frac{W}{\sigma_y}\right)\exp\,(-E/RT) \qquad (10.21)$$

10.4.1 *Example.* Taking

$\beta = 0.33$; $\gamma = 1.2$; $d = 1.13 \times 10^{-7}$ cm; $v = 10$ cm/s; $t_o = 2.9 \times 10^{-12}$s;

$E = 11{,}500$ cal/mole; $R = 1.99$ cal/^{o}K - mole; $T = 311^o$K,

from equation 10.21,

$$\frac{V}{S} = 0.131 \times 10^{-6}\left(\frac{W}{\sigma_y}\right)$$

The volume rate of wear per unit sliding distance can be obtained at any load for a metal, knowing its flow stress. Note that the dimension for the wear rate is cm^3/cm.

REFERENCES

1. Bryant P J, *Trans 9th Nat Vacuum Symposium,* (1962-63), p. 311.
2. Rowe C N, *Trans ASLE,* (1966) *9*, 101.
3. Kingsbury E P, *J1 Appl Physics,* (1958) *29,* 888.
4. Kingsbury E P, *Trans ASLE,* (1960), *3,* 30.
5. Bowden F P, Gergory J N and Tabor D, *Nature (London)*, (1945),*156*, 97.
6. Rabinowicz E, *Proc Phys Soc B,* (1955),*68,* 603.
7. Rabinowicz E and Tabor D, *Proc Roy Soc A,* (1951) *208,* 455.

CHAPTER 11

ABRASIVE WEAR

When a hard body slides over a soft surface, the applied normal stress ploughs a series of grooves in the latter and this is referred to as the two-body abrasive wear[1]. On the other hand, loose hard particles entering the sliding interface would act as grits and the process of metal removal by these has been termed the three-body abrasive wear. The hard particles may be trapped metallic debris as a result of attrition or detached oxide films. A three-body abrasive wear situation gives rise to accelerated wear and, according to Barwell[2], increases the propensity to scuffing of machine parts.

11.1 *Abrasive Wear Coefficient*

An equation to express the rate of wear by abrasion has been derived for the case where a hard rider interacts with a flat soft body. Consider one single conical asperity with a base angle θ penetrating a mass of soft metal as shown in Fig. 11.1. Let the depth of indentation under a normal load ΔW be h. Suppose that the base diameter of that portion of the cone which intersects the surface of the indented soft metal is 2r. That is 2r is the basal diameter of the cone at a distance h = o. Let this cone plough through a distance ds generating a volume dv of the soft metal.

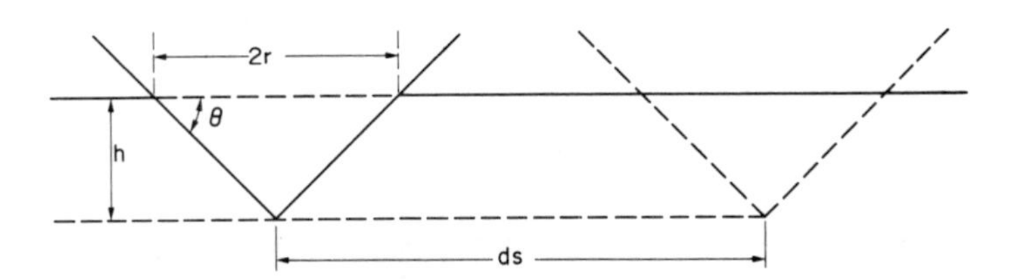

Fig. 11.1 A hard conical asperity ploughing through a soft metal.

As shown in Fig. 11.2, the area of metal displaced in the vertical plane is the area generated by the triangle ABC whose base is the diameter of the basal circle of the cone. Since AB = 2r and the height of the triangle is h, being the same as that of the indented portion of the cone, the projected area of the contacting asperity in the vertical plane is rh.

Therefore, $dv = rh\,ds$

From Fig. 11.1, $h = r \tan\theta$

therefore $dv = r^2 \tan\theta ds$ (11.1)

Since the flow pressure of the soft metal is proportional to its hardness, H, as an approximation

$$\frac{1}{2}\,\pi r^2 = \frac{\Delta W}{H}$$

Thus equation 11.1 becomes

$$\frac{dv}{ds} = \frac{2\Delta W}{\pi H} \tan\theta$$

Adding the contributions by all the asperities of the hard surface in the contact area,

$$\frac{V}{S} = \frac{2W}{H} \frac{\tan\theta}{\pi} \qquad (11.2)$$

All the protuberances on the hard body have been regarded here as having the same angle θ. This is not correct. However, θ in equation 11.2 can be regarded as the arithmetic mean value of the base angles of all the asperities involved in the ploughing of the soft metal.

Comparing equation 11.2 with equation 8.9 which gives the law of adhesive wear,

$\beta = \frac{2\tan\theta}{\pi}$, since $3\sigma_y \simeq H$, so that equation 8.9 is $\frac{V}{S} = \beta\frac{W}{H}$.

If the β term is considered to be equivalent to the wear coefficient β_{ab} under a two-body abrasive situation,

$$\beta_{ab} = 0.63 \tan\theta \qquad (11.3)$$

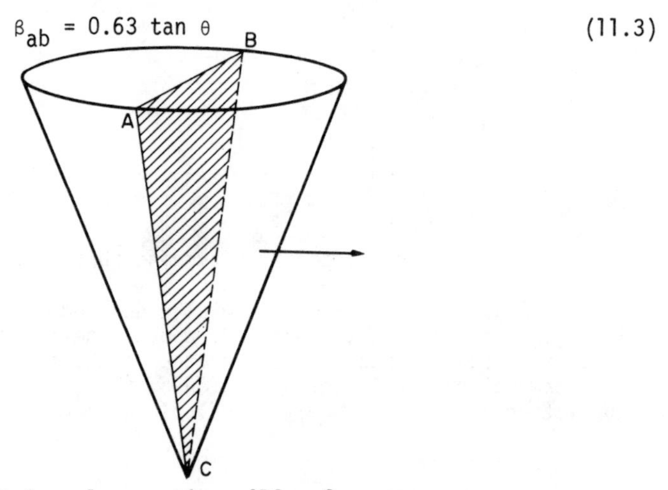

Fig. 11.2 The central triangular section, ABC, of a cone.

11.2 *Abrasive Wear Resistance.*

Bowden and Tabor[3] show that, during ploughing, the total volume of the grooved track per cm of sliding is $W/2\sigma_y$, where W is the applied normal load and σ_y is the flow pressure of the soft material which is being abraded. The abrasive wear resistance R is then expressed as

$$R = 2\sigma_y/W \qquad (11.4)$$

That is, the harder a material is, the more resistant it is to abrasion. Kragelskii[4] has summarised the work of Krushchov and Babichev who have obtained a relationship as follows for commercially pure metals.

$$\frac{R_2}{R_1} = a\,\frac{H}{\sigma} \tag{11.5}$$

where

R_1 = wear resistance of a reference sample

R_2 = wear resistance of the material under investigation

a = constant

H = hardness of the material which is under investigation

σ = apparent normal contact pressure.

It is not clear as to how the wear resistance was measured. However, a plot of R_2/R_1 for a number of metals such as cadmium, aluminium, armco iron, copper, nickel etc. against their respective hardness values shows a linear relationship passing through the origin. A series of heat treated steels also show this linearity although the lines do not pass through the origin but the authors suggest that they will do so if, in each case, the reference sample is the same material in its annealed condition.

11.3 *Abrasives at the Interface.*

Whereas equation 11.2 provides the basis for calculating the rate of wear by abrasion, the situation at the sliding interface is quite complex. Assuming that grit and dust have been excluded from machine parts, the abrasives are the self generated wear debris which should work harden and undergo a size change.

It has been argued[5] that the presence of metallic wear debris at the sliding interface may actually reduce the rate of abrasive wear. It is suggested that particles of debris will adhere to surfaces and undergo comminution until they work harden considerably. The interacting surfaces are then partitioned by an amount of work hardened debris of unspecified composition. The author[5] states that only about 30 per cent of the total reduction in wear during continuous sliding can be attributed to the work hardened layers of the mating surfaces. The remainder is apportioned to the partitioning effect of the wear debris.

Experiments have also been reported[6] on the surfaces sliding when abrasives were introduced at the interface. Systems such as Cu-Zn, Cu-Ni and Cu-Au were abraded and their surface textures studied by electron diffraction technique. It was found that surface orientation after sliding in the presence of abrasives was similar to the rolling textures of metals. In the Cu-Ni system, the coefficient of friction remained sensibly constant up to a certain composition. However, with increased microhardness of the alloy, a concomitant decrease in the rate of wear was observed.

11.4 *Stored Energy.*
Considering again a hard asperity ploughing through a soft metal, the hardness of the latter reaches an equilibrium value as abrasion continues. Deformation of surfaces means also that an amount of energy is stored in the metal as the indenting body passes by.

This idea has been used to deduce a relationship between the wear and the strain energy of metals[7]. It is postulated that an asperity indents into the soft metal and a horizontal pull removes the volume generated by this indented mass and the distance slid. In so doing, the metal removed by abrasion accumulates a maximum amount of stored energy due to mechanical working. Once this equilibrium stage is reached, the strain energy due to further indentation is invariant during subsequent abrasion.

Let for a sliding distance of 1 cm,

μ = measured coefficient of friction,

W = total normal load

g = acceleration due to gravity,

Q = amount of heat evolved due to abrasion,

M = mass wear,

ρ = density of the metal being abraded,

E_m = maximum possible strain energy per unit volume of the soft metal,

E_o = strain energy per unit volume of the soft metal before it has been mechanically worked,

A = total surface area of the groove produced by abrasion for 1 cm sliding distance,

γ = surface energy of the soft metal per unit area.

The work done per cm sliding distance is μWg and this work is expanded in the form of heat liberated, Q, due to friction, the stored energy in the volume of the wear debris produced and the surface energy of the newly created groove. That is,

$$\mu Wg = Q + \frac{M}{\rho}(E_m - E_o) + A\gamma \qquad (11.6)$$

Equation 11.6 can be simplified into the following form:

$$\frac{M}{\rho}(E_m - E_o) = \alpha\mu Wg \qquad (11.7)$$

where α is a constant and is a measure of that fraction of the deformation work which remains as stored energy in the metal. This is found to be small. For example, while deforming a Au - Ag alloy by drawing, 1% of the total deformation energy was found to be stored.

From equation 11.7, the mass wear for 1 cm sliding is

$$M = \frac{\alpha \mu \rho W g}{(E_m - E_o)} \qquad (11.8)$$

For as-cast and annealed metals undergoing abrasion, the term E_0 can be neglected, so that

$$M = \frac{\alpha \mu \rho W g}{E_m} \qquad (11.9)$$

REFERENCES

1. Rabinowicz E, *'Friction and Wear of Materials,'* John Wiley and Sons,(1966).
2. Barwell F T, *Proc Inst Mech Engrs (London),* (1967-68), *182/3K,* 1.
3. Bowden F P and Tabor D, *'The Friction and Lubrication of Solids', Part II,* Oxford University Press, (1964).
4. Kragelskii I V, *'Friction and Wear',*Butterworths, (1965).
5. Sata T, *Wear,* (1960), *3,* 104.
6. Willman H, *Metals and Materials,* (1967),*1,*290.
7. Alison P J, Stroud M F and Willman H, Lubrication and wear third convention. *Proc Inst Mech Engrs (London),* (1964-65), *179,* (3J), 246.

CHAPTER 12

WEAR DEBRIS

Generally speaking, trapped wear debris may score the sliding interfaces of machines although, as summarised, in chapter 11, a reduction in the amount of wear has also been predicted. Whereas friction is expressed as the sum of the resistance to shear of the number of junctions formed at the prevailing normal load, it does not follow that the disruption of the points of contact will simultaneously give rise to wear debris. For the time being, the general laws of wear have accounted for the chance happening of wear debris by incorporating a probability factor and it is implied that this depends on the nature of the metal combination of the couple, the mechanical properties of the member under study and the external variables such as load.

12.1 *Energy Consideration.*

Davies[1] has proposed a model based on energy consideration to explain the mechanism of wear. By implication, it is concluded that the size of a wear debris is that of an asperity which is fixed to the body of the metal at an interface (Fig. 12.1). The energy model postulates that, when two surfaces come into contact, the domains receive energy at the moment of intimacy. The domains are considered to be distributed randomly in space and time. When the surfaces move and the contact is over, the energy received by a domain diffuses into the body of the material according to the equation

$$E = E_0 \exp(-Kt) \qquad (12.1)$$

where E is the energy at a time t

K is a constant and

E is the energy of the body in its ambient condition.

This is a similar idea to that of an abrasive deforming a surface to its maximum potential of strain energy (chapter 11).

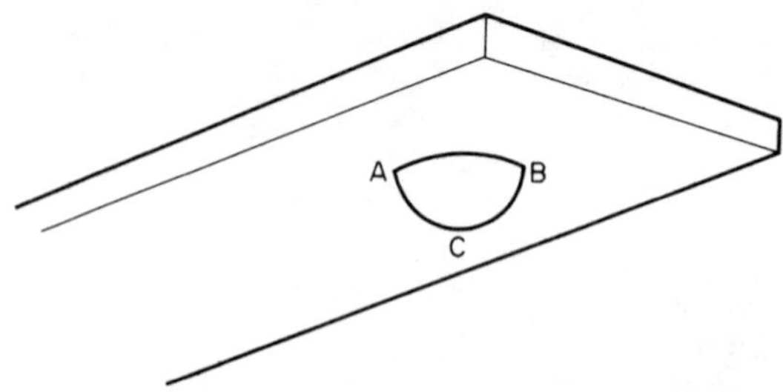

Fig. 12.1 An asperity CAB held to a flat surface forming an interface AB.

Consider the asperity of Fig. 12.1, having OM as the energy of adhesion of it to the substrate (Fig. 12.2). In other words OM is the amount of external energy that must be exceeded before the asperity detaches from the main body of the metal. Consider now the asperity receiving an amount of energy at time t = zero given by the origin O in Fig. 12.2. Suppose the energy received by it at the instant of collision is OA which decays exponentially as shown by the curve AA′. Let the domain receive a second quantum of energy at a time given by B. At this instant, the asperity has an amount of residual energy BB′ left in it so that the result of the second encounter will be for the asperity to possess a total energy given by B′C and BC = OA. At C, the energy decays exponentially as before given by CC′ in Fig. 12.2. Suppose at the third encounter the same amount of energy as at O and B respectively is received at D′ and the energy of the asperity is D′E and at the fourth encounter it is FF′. Since FF′>OM, the asperity detaches giving rise to a wear debris. In other words, an asperity would continue to store elastic energy in itself due to successive mechanical deformation until the total stored energy exceeds the force of adhesion between itself and the substrate when a wear particle is produced.

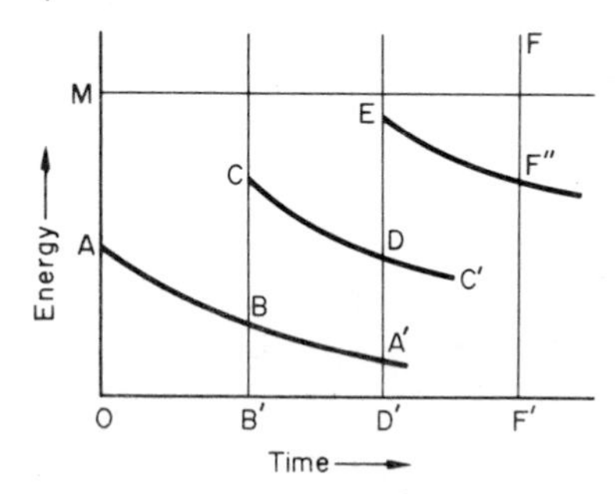

Fig. 12.2 Energy received by an asperity during repeated encounter (idea of Davies[1]).

12.2 *Debris Size*

As a load is applied, junctions form by plastic flow. The interfacial areas of these are not equal, that is the size of any one junction is different to another in the neighbourhood. It is shown[2] that, at an interface, the smaller junctions are the more numerous but they can grow in size upon increasing the normal load, or possibly as a result of a tangential pull. A junction remains attached to the bulk until such time as it grows into a critical size when metal fragments are detached forming loose wear debris.

Consider again an asperity attached to a substrate. An order of magnitude assumption is that once this asperity has been contacted by the other surface and after the latter has moved on, the residual energy in the particle is about 10% of the energy of adhesion. If V is the volume of the asperity and σ_y and E are its yield pressure and Young's modulus respectively, the total energy of it is

$$\frac{1}{2}\ \frac{\sigma_y^2}{E}\ V.$$

Therefore the residual energy after an encounter is $\frac{1}{10} \times \frac{1}{2} \times \frac{\sigma_y^2}{E}\ V$ (12.2)

If W is the work of adhesion and A the area at the interface, the adhesive energy of the asperity which keeps it attached to the main body of the material is WA.

Assume the asperity to be hemispherical with a base diameter 2r. Then,

$$V = \frac{2}{3}\pi r^3 \quad \text{and}$$

$$A = \pi r^2$$

A loose wear fragment will form when, for the asperity,

Elastic stored energy ⩾ energy of adhesion (12.3)

The elastic stored energy is given by equation 12.2. Substituting the values of V and A in terms of r, the condition 12.3 can be expressed as

$$\frac{1}{20}\frac{\sigma_y^2}{E}\frac{2}{3}\pi r^3 \geqslant W\pi r^2$$

that is

$$r \geqslant \frac{30EW}{\sigma_y^2}$$

or

$$2r \geqslant \frac{60EW}{\sigma_y^2} \qquad (12.4)$$

Making an assumption that, for most metals,

$\sigma_y = \frac{H}{3}$ and $\frac{\sigma_y}{E} = 3 \times 10^{-3}$ where H is the hardness and

substituting in equation 12.4,

$$2r \geqslant 60\ 000\ (W/H) \qquad (12.5)$$

Rabinowicz[3] has observed that, in wear experiments with steel, the critical size of a fragment was 12×10^{-3} cm before it was detached.

The quantity W/H in equation 12.5 is a measure of the ratio of surface to volume energies of a fragment and is regarded as important in wear studies. Hard metals have high strengths and usually low surface energies so that a small critical diameter, 2r in equation 12.5, is expected. A practical useful effect should be that run-in surfaces will be smooth for hard metals. On the other hand, the wear particle size is large with softer surfaces because of a high W/H ratio, resulting in a rough finish during sliding. These deductions have been confirmed experimentally[4].

12.3 *Effect of Load*

If a metal such as copper is loaded against steel an amount of the former is transferred on to the latter. In this way the area of the junction could be estimated by, for example, autoradiographic technique. Area and volume of wear debris formed during shearing of junctions are not altogether easy to evaluate experimentally but Rabinowicz[2] has shown that the mass M of the largest fraction of wear debris could be related to the applied normal load

W as

$$M = CW^{\alpha} \tag{12.6}$$

where C is a constant and α, for copper on steel, has a value of 0.3. The fractional nature of α in equation 12.6 shows that the mass of the wear fragment only increases slowly as the load is increased. The same study[2] showed that the predominant effect of increasing the applied load was to increase the number of junctions.

REFERENCES

1. Davies R, *'Friction and Wear', p.1,* Elsevier, (1959).
2. Rabinowicz E, *Proc Phys Soc B,* (1953), *66,* 929.
3. Rabinowicz E, *Wear,* (1958-59), *2*, 4.
4. Rabinowicz E and Foster R G, *Amer Soc Mech Engrs, Lubrication Symposium, Boston, Mass., June 2-5,* (1963), *Paper No. 63 - Lub S-1.*

CHAPTER 13

METAL TRANSFER

It is evident that the process of wear under sliding situations is associated with the formation and growth of junctions nucleated by contacting asperities. Plastic interaction is also responsible for frictional resistance. It is surprising, therefore, that a decisive correlation between friction and wear is not generally obtained. It is probable that sliding under load causes a change in the metallurgical properties of the couple and the effect has a significant influence on the mode of wear. There would appear to be a dearth of information in published literature with regard to the metallurgical changes in solids in terms of friction and wear. Fortunately, interest in the role of solid solubility, crystal structure, microhardness and surface textures of sliding surfaces has grown and the next decade should produce a body of knowledge on the metallurgy of friction and wear generally.

13.1 *Steel on Brass*

When a steel shaft slides on a brass bearing, the first effect is for an amount of brass to be deposited onto the steel. With continued sliding, the deposit flakes off into wear debris followed by a further transfer of the soft metal on the shaft. The alternate deposition and disruption continues as long as there is wear of the bearing. This is basic in wear processes in metals and has been the subject for detailed study by a few workers.

Quantitative studies of metal transfer have been undertaken by radioactive tracer technique in which an irradiated pin was slid on an inactive surface and the amount of metal transfer was recorded by a calibrated Geiger-Muller counter attached to the wear machine. Kerridge and Lancaster[1] have followed the progress of 2 per cent leaded 60/40 brass pins sliding on hard steel. The nature of transfer and wear of brass with time as shown in Fig. 13.1 shows that the first effect of sliding is for the brass to be transferred rapidly onto the stellite ring so that the pin wears at a fast rate. However, a stage is reached when the amount of metal transferred reaches a constant value when the rate of wear assumes a steady state. Figure 13.1 also shows that, even at a very high load of 22.5 kg, under lubricated condition the amount of metal transferred is small and the rate of metal removal per unit time is the same as that observed when the load was much lower but the interface was dry. The results from these experiments show that the wear of brass on steel occurs at two distinct stages.

As sliding commences, the immediate effect is for the pin to deposit an amount of brass onto the steel and there is an absence of loose wear debris. The amount of metal transfer increases exponentially with time and continues (Fig. 13.1) for about 2.75 min and 5 min for the dry and lubricated conditions respectively. This is followed by the second stage when loose wear debris form from the transferred layer of the brass.

Continued sliding results in an increase in the size of the fragment and in the surface roughness of the deposited brass. Metal is transferred in the form of discrete particles and the authors show that a wear fragment detached from the interface is about eight times the area of the transferred particle and six times its thickness.

The phenomenology of transfer and the subsequent detachment of brass can be described in the following way. Suppose (Fig. 13.2) a particle of brass is transferred onto the second surface during the first effective encounter and the fragment has a diameter a_1 and height h_1. Further transfer occurs at a future encounter on the first deposit but both the diameter and height of the new deposit have changed to a_2 and h_2 respectively. By observation, $a_2>a_1$ and $h_2>h_1$. This way, the first fragment grows by successive deposition of metal until its diameter is a_n and height h_n, where $a_n>...>a_4>a_3>a_2>a_1$ and $h_n>...>h_4>h_3>h_2>h_1$. This critical diameter a_n is reached after about 50 encounters and the particle is detached.

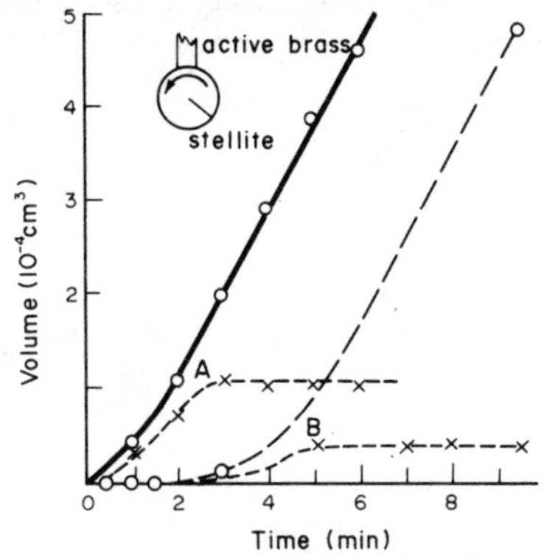

Fig. 13.1 Variation of wear and metal transfer of a brass pin sliding on stellite ring with time. -o wear; ----x---- transfer. A, unlubricated, load 5 kg; B, lubricated by cetane, load 22.5 kg. (Kerridge and Lancaster[1]).

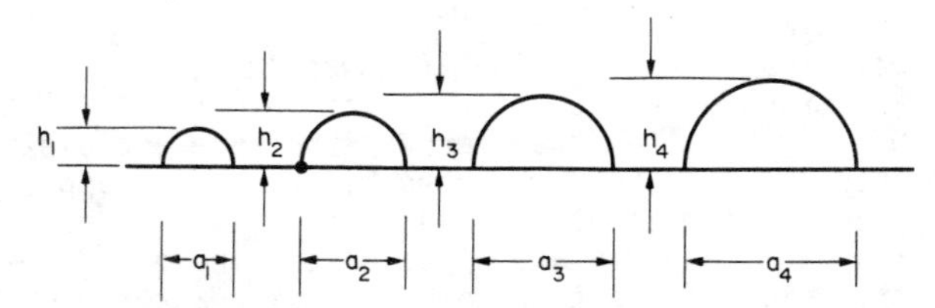

Fig. 13.2 Increase in the diameter and height of a transferred fragment with time due to repeated encounters.

13.2 *Steel on Steel*

It was observed[2] that, when annealed tool steel slid on hard tool steel, the transferred film of metal oxidised before being detached and the rate of production of wear debris was determined by the oxidation rate of the steel. This is different to the wear of brass on steel where the speed with which loose debris were produced was shown to be governed by the frequency of metal transfer from the pin to the disc. However, for the steel couple, formation of wear debris is still a multistage process. The stage by stage mode of wear in metals has also been confirmed by Archard and Hirst[3]. Tool steel pins were slid on rings of the same material and the progress of wear was ascertained by radioactive tracer technique, followed by optical and electron microscopic examination of surfaces. With the onset of sliding, the pin lost just enough weight to establish geometric conformity with the disc. After this, there was a high rate of metal transfer from the pin to the disc, the density of the deposit increasing with sliding distance. A period of transition followed when the pin bedded in over the full width of the track and there was a fall in the density of the transferred layer of metal from the soft pin to the hard disc. During this stage, the pin wore at a high rate and this was followed by an appreciable loss of metal from the ring. The system then reached equilibrium when both the rate of metal transfer and wear reached a constant value. This equilibrium stage was reached after 10,000 to 20,000 revolutions and the rates of wear thereafter of the pin and ring were of the same order of magnitude.

It appeared that the initial debris formed by detachment of the already deposited metal on the ring. The particles were large because, at this stage, the normal load was sustained by a relatively few contact points. As sliding prolonged, the junctions grew and metal transfer and wear continued but the debris size got progressively smaller. In the final stage, a different mechanism seemed operative. Thus, although metal transfer still took place, the wear process appeared to be largely independent of this, and the height of the fragments were greater than those of the underlying steel asperities. The detritus comprised aggregates of particles from 100Å in size and showed the presence of oxides. The authors[3] concluded that there were two basic mechanisms of wear viz.,

(1) metal transfer followed by abrasion,

(2) a single stage process of direct abrasion.

Both these mechanisms could operate simultaneously over the whole interacting interface.

13.3 *Amount of Transfer*

Irradiating a specimen and then sliding it on an inactive surface to estimate the amount of metal transfer was carried out by Sakmann et al[4] and, although this gives the quantity of metal transferred, the size and distribution of these deposited fragments are best measured by autoradiographic technique. In this, a photographic film is placed over the surface after sliding the radioactive metal over it and the amount of blackening gives a measure of the quantity and distribution of the transferred metal. Experiments[5] show that the particles appear as discrete patches and not as a continuous layer over the surface. Work carried out[6] with various metal combinations has established that the amount of metal picked up during sliding is the same order of magnitude irrespective of the hardness of the material. In the absence of a lubricant, the amount of pick up is about 40 times more for similar metal combinations than for couples of dissimilar compositions. Studies with lubricated systems show that the amount of metal transfer can be reduced by a factor of 2×10^4 or more. The mode of transfer during sliding, however, is the same in that the transfer of metal takes place as discrete patches. It was also noted that over a load range of up to 4 kg, for metals such as copper, platinum or zinc, the amount of pick up was proportional to the load but surface roughness or speed in the range 0.003 - 0.4 cm/s had little effect.

REFERENCES

1. Kerridge M and Lancaster J K, *Proc Roy Soc A*, (1956) *236*, 250.
2. Kerridge M, *Proc Phys Soc B*, (1955) *68*, 400.
3. Archard J F and Hirst W, *Proc Roy Soc A*, (1957), *238*, 515.
4. Sakmann B W, Burwell J T and Irvine J W, *Jl Appl Physics*,(1944), *15*, 459.
5. Gregory J N, *Nature*, (1946), *157*, 443.
6. Rabinowicz E and Tabor D, *Proc Roy Soc A*, (1951),*208*, 455.

CHAPTER 14

SURFACE AND SUBSURFACE

The importance of careful preparation of surfaces involved in the kinematic chains of machinery is well appreciated as exemplified by the caution exercised in the running-in of engines. A well polished surface suffers less damage in service[1] so that it would appear that running-in involves close conformation of opposing surfaces. Running-in also means that both the surfaces and the underlying layers in the components must undergo an equilibrium state of work hardening, although none of these factors appear to have been proved by definitive experiments.

14.1 *Surface Layers and Sublayers*

The nature of mechanically polished surfaces has been studied by optical and electron microscopy, by the use of stereoscan and electron diffraction technique or by electrical contact resistance measurements[2-4]. The latter technique can give an assessment of the degree of oxidation of surfaces whereas metallography can reveal the nature of the layers and sublayers on sectioned specimens.

Structural changes due to sliding of steel against itself have been studied[5,6] with the aid of electron microscope and by x-ray analysis. Two distinct layers from the sliding interface have been observed. The layer adjacent to the surface showed a high dislocation density as is to be expected of a heavily deformed metal. Further below towards the body of the parent metal was a second layer which was less severely deformed. The boundary between these layers became more diffuse as the carbon content of the steel decreased.

14.2 *Friction*

Effect of the degree of work hardening on the wear behaviour of fcc metals such as aluminium, copper, silver and gold has been studied[7]. Annealed, 2.0 x 1.5 x 0.6 cm thick rectangular blocks of these metals were abraded on steel files under a load of 1 kg at sliding speeds in the range 1 - 5 cm/s. The aluminium sample developed a work hardened layer, being as hard as cold rolled aluminium reduced by 80%. The coefficient of friction μ could be related to the surface microhardness H by

$$\mu = \ell - mH \qquad (14.1)$$

where ℓ and m are constants.

Generally, work hardening resulted in diminished rates of wear.

14.3 *Surface Fatigue*

Surface fatigue may play a significant role in producing wear debris as seen from the experiments of Bayer and Schumacher[8]. A hard steel sphere was slid under a reciprocating condition on single crystals of copper and observations were confined to the copper only, whose sliding faces were [100], [111] and [225] planes respectively. Two main types of experiments were carried out.

In one group, the load was varied but the run comprised one single stroke only. In the second series, the load was kept constant while the total sliding varied up to 10,000 strokes.

There was no significant change in surface topography between the various planes upon sliding. However, faint striations appeared in the sliding direction for any surface at low stresses, being pronounced as the load increased. When the applied pressure reached a value of about one half the yield stress of copper in shear, surface damage in a plane perpendicular to the sliding direction appeared. Transverse sections of specimens revealed that these were in the form of cracks, being referred to as cross-hatching by the authors. As the applied stress was further increased, cracks and spalling of surfaces were noticed and the copper flowed plastically. Gross plastic flow and slip bands were evident as the load was increased beyond the yield point of copper in shear. The effect of repeated sliding at a load of half the yield shear stress of the material was to produce a similar sequence of change in the surface. The fact that even at low stresses, the sequence of events as striations, cross hatching, cracks and spalling led the authors[8] to conclude that surface fatigue played the leading part in producing wear debris.

14.4 *Plasticity Index*

Both the theories of friction and wear assume plastic condition of the sliding interface. This must be so as a machine is started for the first time but metallurgical examinations show that the interface work hardens to a finite depth. If a surface is hard, further plastic flow should not be expected.

This has been discussed in detail by Greenwood and Williamson[9] and they have used the concept of the plasticity index ψ which is essentially the ratio of the hardness of the material before it is slid under load to that hardness obtained after sliding. The authors show that the plasticity index is greater than unity when conditions are plastic but the situation at the interface is elastic when $\psi < 0.6$. The plasticity index is defined as

$$\psi = \frac{E}{H} \left(\frac{\sigma}{R}\right)^{\frac{1}{2}} \tag{14.2}$$

where σ is the standard deviation of asperity heights and R is the mean asperity radius of curvature. Expressions for the plasticity index have also been given by Blok[10], Halliday[11], and recently by Halling et al[12] who use measurable parameters and show[12] that

$$\psi = \frac{E}{H} n \sin \theta \tag{14.3}$$

where E = modulus of elasticity of the material,

H = hardness of the material,

θ = base angle of an asperity, assuming it to be conical

and n has a value between 1.7 - 2.4.

The conclusion of Greenwood[13] from theories of surface topography is that the distinctiveness of plastic flow in dynamic mechanical situations is much exaggerated. Plasticity index value is the important quantity predicting the nature of contact. Recent theories emphasise the role of surface texture parameters such as surface density, height distribution and the radii of curvature of asperities on the nature of surface interactions. Williamson[14] has plotted what he calls microcartography of surfaces based on digital analysis of results obtained from profilometric measurements. These maps show individual features of surface texture over a large area and the technique should be valuable to research workers in view of the current emphasis on the nature of surface topography with regard to the mechanism of friction and wear.

REFERENCES

1. Moore W H, *Product Engineering,* (1958),*29,*(48), 63.
2. Matsunaga M, *Report of the Institute of Industrial Science, University of Tokyo,* (1958), *7,* (5).
3. Quinn T F J, *'The Application of Modern Physical Techniques to Tribology,'* Newnes-Butterworths, (1971).
4. Cochrane W, *Proc Roy Soc A,* (1938), *166,* 228.
5. Nakajima K and Kawamoto J, *Wear,* (1968), *11,* 21.
6. Nakajima K and Mizutani Y, *Wear,* (1969), *13,* 283.
7. Lin D S, *Wear,* (1969), *13,* 91.
8. Bayer R G and Schumacher R A, *Wear,* (1968), *12,* 173.
9. Greenwood J A and Williamson J B P, *Proc Roy Soc A,* (1966), *295,* 300.
10. Blok H, *Proc Roy Soc A,* (1931), *171,* 79.
11. Halliday J S, *Proc Inst Mech Engrs,* (1955),*169,* 777.
12. Halling J and El-Refaie M, *Tribology Convention (1971), 12-15th May, Inst Mech Engrs, P.60.*
13. Greenwood J A, *Amer Soc Mech Engrs,* (1965), Paper No. 65 - Lub - 10.
14. Williamson J B P, *Proc Inst Mech Engrs,* (1967-68), *182,* 21.

CHAPTER 15

TEMPERATURE AND SPEED

Friction and wear of metals are shown to be governed largely by the interaction of asperities of two sliding surfaces. Energy dissipated due to mechanical work inevitably causes a rise in temperature but it happens intermittently in so far as the points of actual contact are concerned due to sticking and then slipping of the junctions. These temperature flashes are short lived in the order of 10^{-4} second. The heat evolved in friction is dissipated to the surrounding with the result that the asperity tips are at a high temperature but the bulk of the component remains relatively cool. An increase in load or speed raises the temperature of the junctions and, in extreme cases, may cause incipient fusion. There is no simple way of measuring the temperatures of the actual contact areas but the bulk temperature of an interacting couple can be obtained. Since both adhesive and abrasive wear laws incorporate the mechanical properties such as yield stress and hardness of the metal and since both these are affected by the thermal environment, a study of friction and wear at high temperatures is of value. It is also important because rotating parts are required to operate at elevated temperatures.

15.1 *Temperature*

A very useful experiment has been carried out by Lancaster[1] who slid 60/40 brass pins on hard steel rings in room temperature while the couple were heated by a surrounding furnace. As the temperature was raised, the effect was to increase the rate of wear of the pin which reached a peak at a characteristic temperature depending upon the normal load (Fig. 15.1). Beyond this peak, there was a fall in the wear rate of the brass. Figure 15.1 shows the rate of wear at a constant load of 2 kg, the ring rotating at a surface speed of 1.3 cm/s. A low surface speed was employed to keep the frictional heating to a minimum.

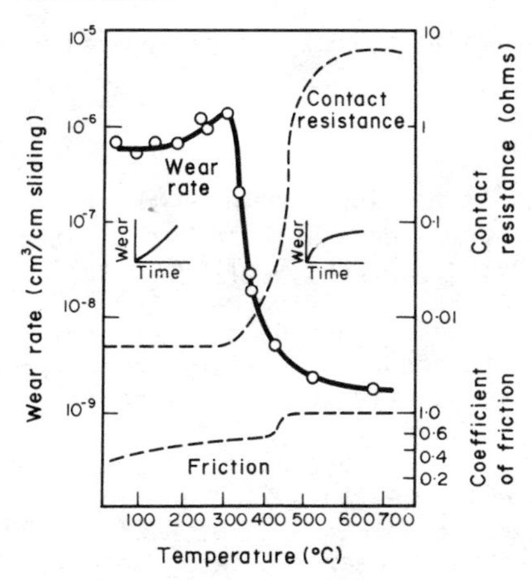

Fig. 15.1 Variation of wear rate with temperature for 60/40 brass on tool steel (Lancaster[1]).

The first regime of wear (Fig. 15.1) shows that the contact resistance is nearly constant but there is a small progressive increase in friction. It is shown[1] that, since the hardness of the brass pin decreases with temperature, the rate of wear, being inversely proportional to this mechanical property, increases. An increase in the contact resistance means a decrease in electrical conductivity. This can happen if the metal junctions are diluted by an insulating material such as oxide, the formation of which is facilitated because of the high temperature of the environment. The reason for a sudden rise in the coefficient of friction is not given.

At any temperature, there is a transition load below which the surface oxides are not completely destroyed and the wear rate is low. Above the transition load, metal to metal contact is extensive and wear is high. The magnitude of the transition load increases with temperature, being 0.2 kg under ambient condition and 6 kg at 350°C. The authors show that the mechanism of wear at any load or temperature is by transfer of brass onto the steel ring and the subsequent detachment of this deposited layer.

Hughes and Spurr[2] have slid wax pins on cast iron discs over a speed range of 100 - 500 cm/s at varying loads from 0.8 - 3 kg. An increase in load and speed caused the temperature to rise and hence the hardness of the pin to fall. The authors measured the hardness of the wax at the resultant temperature given by a particular combination of load and speed. If the rate of wear is now plotted against the reciprocal of this measured hardness, a linear relationship is obtained as shown in Fig. 15.2.

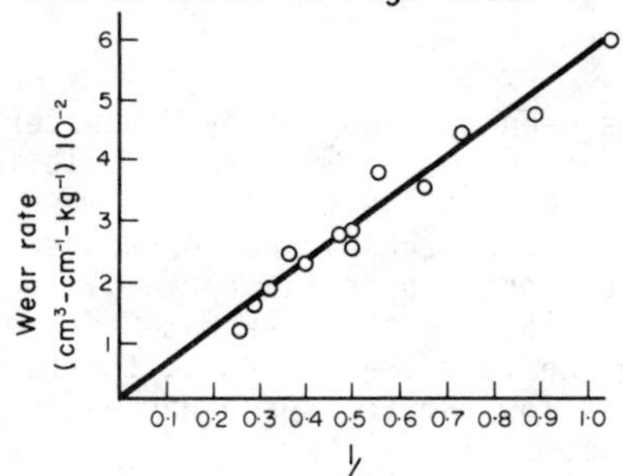

Fig. 15.2 A plot of wear rate against the reciprocal of hardness (Hughes and Spurr[2]).

15.2 *Speed*

Since the effect of speed is to cause an increase in the temperature of the sliding interface, oxide formation should be facilitated. An increase in temperature also means that the hardness of the metal will decrease so that an increase in the rate of wear should be expected. This is so if the interface temperature is high enough. Indeed, at very high speeds, melting of the rubbing surfaces is known to occur, the effect being accentuated if the melting point and the thermal conductivity of the material is low.[3]

However, the general effect of increasing surface velocity is to cause a reduction in the rate of wear[4] and this is shown in a detailed study of wear of 60/40 brass on steel over a speed range of 0·01 to 500 cm/s using a pin-ring machine[5]. The variation of wear rate at a load of 22.5 kg with speed is shown in Fig. 15.3 which shows a fall followed by a rise in the rate of metal loss beyond a speed of about 100 cm/s. The authors[5] observed the rate of metal transfer by using radioactive tracer technique and Fig. 15.3 shows that this is the same as the rate of wear. In other words, the basic mechanism of wear remains unaltered irrespective of the surface speed, that is metal is first transferred to the bush from the pin and wear debris is produced from this deposited layer. To see if it was the effect of a rise in temperature, a set of experiments were conducted where the pin was water cooled. Figure 15.4 shows that in that case the rate of wear continued to fall. Conversely, if the brass was thermally insulated from the pin holder, the increase in wear at the high speed range was greater than that obtained when the pin was not insulated or was water cooled.

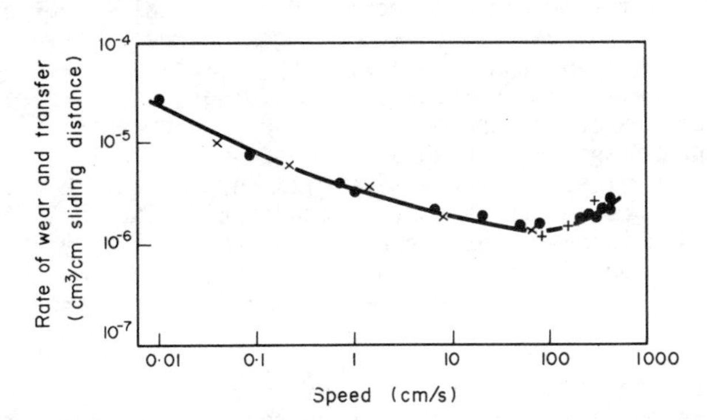

Fig. 15.3 Variation of wear rates and metal transfer with speed. Load 22.5 kg. ●, Rate of wear; x, rate of transfer below 65 cm/s; + rate of transfer above 65 cm/s. (Hirst and Lancaster[5]).

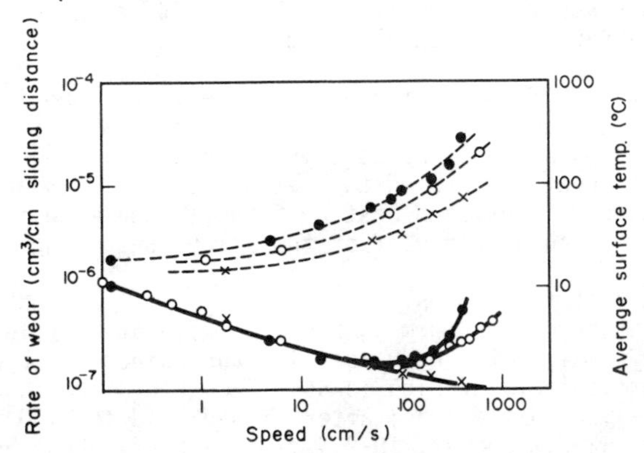

Fig. 15.4 Variation of wear rate and surface temperature with speed for thermally insulated and cooled pins. Load, 3 kg. ——wear rate; ----surface temperature; O, metal pin-holder; x, water-cooled pin holder; ●, thermally insulated pin-holder. (Hirst and Lancaster[5]).

Figure 15.4 shows that the temperature, as recorded by a thermocouple 2 mm behind the interface, increases progressively with speed and there is no abrupt rise in temperature at the speed where the rate of wear begins to increase. Calculations suggest that for the watercooled pins, the flash temperature at the contact points can be as high as 260°C when the temperature recorded by the thermocouple is 70°C. These flash temperatures are quite high and there is no sudden rise in temperature at the point of deviation of the rate of wear (Fig. 15.4). The authors[5] therefore discount any direct effect of surface temperature on the mode of wear but suggest that wear rates of brass are governed by two sets of factors, viz.,

(1) those which influence the size of the individual transferred fragment,

(2) those which determine how frequently they are formed.

Observations in these experiments show that the size of the transferred fragment decreases as the speed is increased as it should be since time is not available for junction growth. Transfer of brass on steel occurs only when the surface of the former has been deformed a critical number of times. This critical number is dependent on speed, being high when the velocity of sliding is increased. This means that the frequency of metal transfer will decrease with increasing surface speed resulting in a progressive fall in the rate of wear. The second effect of speed is to cause a progressive rise in temperature with a concomitant fall in shear strength of the brass, resulting in an increased junction size. A progressive rise in temperature with increased speed, therefore, causes the amount of brass which is deposited to get larger and, when this overcomes the effect of the frequency of transfer, wear rate increases. Figure 15.4 shows that the rate of wear of the thermally insulated pin rises at a lower surface speed and is also greater than the pins which are cooler.

Cocks[6] attributes the reduced wear rate of metals to the protective effect of oxide films when surfaces interact at high speeds. Several metal combinations were studied at varying loads up to 3 kg on an apparent contact area of 0.36 cm^2. The surface speed in these experiments was 6500 cm/s and several couples of like metals such as copper and nickel were used. The other combinations were copper-steel and tungsten carbide sliding on itself. Low rates of wear were observed for all couples except nickel on nickel.

Diminished wear rates of copper-tin alloys have been found to be dependent on the amount of tin and this would appear to be due to an alloying effect but oxidation is observed as well, especially at high sliding velocities. Experiments of DeGee et al[7] showed that at a load of 10 kg and a surface speed of 100 cm/s, for test runs each lasting 2 hours, the weight loss of a 12% tin-copper alloy was about one half of the alloy containing 5% tin.

It was observed[8] that the general effect of sliding steel on bronze was the pick-up of the latter by the former, particularly at high loads. Frictional heating occurred at high sliding velocities and oxides, probably of copper, were apparent particularly at high loads. Electron microprobe analysis of the transferred layer on the steel after it had slid for 100 m showed the composition to be 62-74% copper, 10-25% iron, 5-6% tin and about 10% oxygen. The bronze itself had 10.6% tin and 0.6% phosphorus. The total wear at a velocity of 100 cm/s was about ¼ of that at 7 cm/s and the amount of bronze transferred to the steel increased with sliding velocity. As in the case of brass on steel, the transferred material was harder than the original metal. Films of oxide formed at the sliding interface and a reduction in the rate of wear was noted.

REFERENCES

1. Lancaster J K, *Proc Phys Soc B*, (1957), *70*, 112.
2. Hughes G and Spurr R T, *Proc Phys Soc B*, (1955), *68*, 106.
3. Bailey A I, *J1 Appl Physics*, (1961), *32*, 1430.

4. Steijn R P, *Trans Amer Soc Mech Engrs D*, (1959), *81*, 56.
5. Hirst W and Lancaster J K, *Proc Roy Soc A*, (1960), *259*, 228.
6. Cocks M, *Paper at Lubrication Conference held in Toronto, Canada, October* (1957), Amer Soc Lub Engrs.
7. DeGee A W J, Vaessen G H G and Begelinger A, *Trans Amer Soc Lub Engrs*, (1969), *12*, 44.
8. Wellinger K, Uetz H and Komai T, *Wear*, (1969), *14*, 3.

CHAPTER 16

SOLUBILITY

Tribological studies, by the very nature of the subject, must be interdisciplinary in character but metallurgically orientated examinations such as those described previously would help towards a more complete understanding of the mechanism of wear. Apart from the micro and macro effects of the interface and the subsurface, the role of solubility and crystal structure is basic in promoting interaction on an atomic scale in wear situations and the present state of knowledge is summarised in this chapter.

16.1 *Solubility*

An increase in the magnitude of friction and wear is experienced when materials slide in high vacuum. The reason for this high friction is attributed to the formation of strong junctions which grow and cohere due to solid phase welding of the asperities. Considering the transient nature of a junction life, at a first glance it is hard to visualise how fresh junctions form and grow, since growth should occur by diffusion on an atomic scale and the process is time dependent. Probably, it is the flash temperatures of the asperities which contribute to junction growth, since diffusion rate is accelerated with increasing temperature.

16.1.1 *Score Resistance*. **Diminished** junction growth and hence score or wear resistance of metals has been correlated with the solid solubilities of metal combinations quite clearly, albeit in a qualitative way[1]. For example lead, tin and indium have very limited solubility in iron and a combination of these three elements with iron provides the basis for a good bearing situation. On the other hand, very poor score resistance is experienced with a combination of iron and nickel which have very large mutual solid solubility.

The score resistance of some 38 different metals against steel was studied[2] by rubbing 1.56 cm square specimens against steel discs rotating at 23.3 m/s, the load on the specimens being varied up to 540 kg. The load bearing capacity of the specimens without seizure was taken as the criterion for score resistance. The performance of the specimens was classified as follows, the maximum loads being reached monotonically to the values stated:

(1) very poor score resistance - seizure when a load of 135 kg was reached;

(2) poor score resistance - when a load of 135 kg could not be sustained for 1 minute of sliding;

(3) fair score resistance - same as in (2) but when the load was increased to 225 kg;

(4) good score resistance - no seizure when sliding could be sustained up to 1 minute at a normal load of 225 kg.

The score resistance of materials according to this classification was as follows:

(1) Very poor - Beryllium, Silicon, Calcium, Titanium, Chromium, Iron, Cobalt, Nickel, Zirconium, Columbium, Molybdenum, Rhodium, Palladium, Cerium, Tantalum, Iridium, Platinum, Gold, Thorium, Uranium.

(2) Poor - Magnesium, Aluminium, Copper, Zinc, Barium, Tungsten.

(3) Fair - Carbon, Copper, Selenium, Cadmium, Tellurium.

(4) Good - Germanium, Silver, Cadmium, Indium, Tin, Antimony, Thallium, Lead, Bismuth.

It is seen from these results that the materials which have good resistance to surface damage have either a limited solubility in iron or form intermetallic compounds with it. Cadmium and copper behaved in an inconsistent way. However, it is proved that the B - subgroup metals show the best resistance to scoring and these elements degenerate progressively into non-metallic type materials along the periodic table. For example, copper, gold and silver have metallic characteristics and are capable of forming strong interatomic bonds, whereas indium, bismuth, lead and others which form intermetallic compounds with iron show covalent brittle bonding. Calcium and barium are immiscible in iron. The reason for their poor score resistance is given to be due to formation of metallic bonds although the junctions do not grow effectively. This explanation is hard to reconcile with the fact that junctions between dissimilar materials can form and grow only when they are soluble on an atomic scale and lack of junction growth means antiscoring property according to this mutual solubility concept.

The authors extended the work[3] to study the nature of surface damage and friction characteristics of these materials against iron. In these experiments hemispherical riders in armco iron were made to slide on flat plates of the elements under examination, polished on emery paper, the direction of sliding being across the scratches produced during surface preparation. Surface damage was assessed by comparing photomicrographs and transfer of iron from the rider on a plate was examined by electrographic spot testing. Although there was no obvious correlation between the coefficient of friction and the degree of solubility between the rider and the flat plate, it was observed that those which were immiscible or the combinations which formed intermetallic compounds showed shallow depths of surface damage. Unfortunately, the authors have not made any attempt to express the rate of wear quantitatively. It would have been useful to carry out detailed examination of surfaces and the layers underneath.

16.1.2 *Work in Vacuum.* Using commercially pure metals, Coffin has examined the solubility aspect by using a pin-ring machine[4]. The disc, 5.62 cm in diameter, constituted one of the metals of the couple to be tested and could attain a maximum surface speed of 0.5 cm/s. The normal load on the rider constituting the other member of the couple was 270 g. The apparatus was designed to work at very low pressures or at chosen atmospheres. The author[4] arrives at similar conclusions as the other workers that mutually soluble metals form coherent junctions and wear easily.

16.1.3 *Cutting the Sliding Surface.* In order that a contaminant or oxide film does not partition the sliding interface, it is necessary to work in high vacuum which is expensive. An ingenious method is that used by Halling[5] who mounted a metal cylinder between the two centres of a lathe. The load bar was so designed that the rider could traverse the cylinder as the lathe saddle moved. A cutting tool cut the surface of the cylinder and a rider wore on this freshly prepared surface. A series of runs were made so that the time interval between cutting and rubbing could be extended. Cutting a surface assumes that the metal is free from contamination. However, with time, the same surface will be contaminated. As expected, it was observed that as the time interval between cutting and rubbing was extended, the rate of wear diminished. However, although relatively inexpensive, the experimental set up is not as easy as it appears to be. Apart from the rigidity the machine must have, consistent surfaces have to be produced by using the correct surface speed and depth of cut. Certain metals by their very nature are difficult to machine, e.g., aluminium.

A similar approach has been used by Sata[6]. Rods, 4 mm in diameter of copper, plain carbon steel, aluminium, brass and some other metals were rubbed respectively on the rim of steel bushes 75 mm od with a 5 mm wall thickness. A cutting tool was placed at the opposite end of the rod so that a freshly created surface of the bush could be exposed to the rubbing test piece at each revolution. The wear continued to increase with sliding distance linearly (Fig. 16.1) when rubbing was on the machined surface. When the cutting tool was omitted a curvilinear running-in wear was followed by a steady state wear (Fig. 16.2).

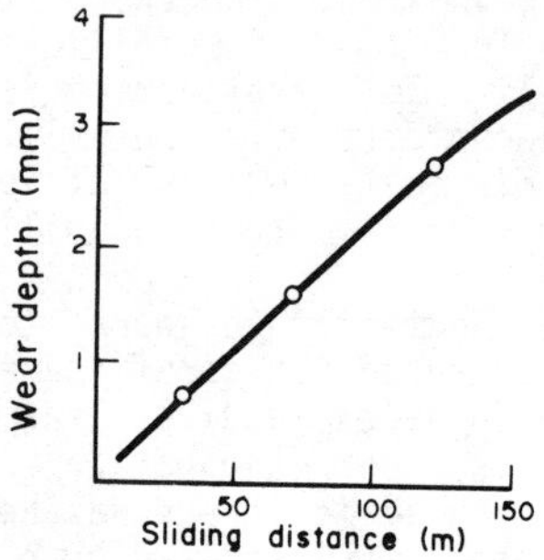

Fig. 16.1 Height loss of copper pins slidjng on steel at an apparent contact pressure of 32 kg/cm^2 and surface speed of 7.5 cm/s. Steel bush cut ahead of the copper pin with a cutting tool (Sata[6]).

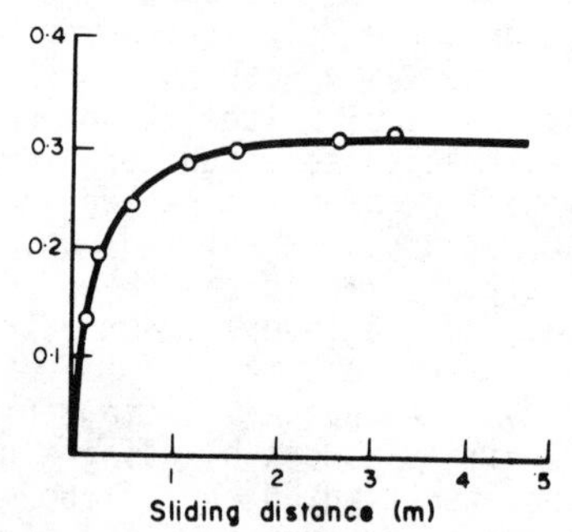

Fig. 16.2 Same situation as Fig. 16.1 except that the steel bush was not cut with a cutting tool. (Sata[6]).

These experiments suggest that contact between surfaces without contaminants is important if the fundamentals of metallic interaction is to be understood.

REFERENCES

1. Roach A E, Goodzeit C L and Totta P A, *Nature*, (1953), *172*, 301.
2. Roach A E, Goodzeit C L and Hunnicutt R P, *Trans Amer Soc Mech Engrs*, (1956), *78*, 1659.
3. Goodzeit C L, Hunnicutt R P and Roach A E, *Trans Amer Soc Mech Engrs*, (1956), *78*, 1669.
4. Coffin L F, *Lubrication Engineering*, (1956), *12*, 50.
5. Halling J, *Wear*, (1961), *4*, 22.
6. Sata T, *Wear*, (1960), *3*, 104.

CHAPTER 17

CRYSTAL STRUCTURE

The phenomena of friction and wear have been attributed to plastic interaction of metals at some stage of the life of a couple. Plastic deformation of metals occurs by slip, i.e. by shearing of planes of atoms over one another. Slip is anisotropic, the direction being almost always the one along which the atoms are closely packed. The slip planes are also those possessing the highest number of metal atoms. Thus the direction of slip varies according to the crystal structure, being < 110> for fcc metals. The bcc metals slip along < 111>, while the direction for C P Hex materials is $(2\bar{1}\bar{1}0)$. The planes of slip for fcc and C P Hex metals are {111} and (0001) respectively. In the bcc structure, the slip planes are {110} , {112} and {123} , all of which being densely packed. The slip planes may be influenced by temperature. For example, aluminium slips on {100} and other planes at elevated temperatures and similarly C P Hex on $\{10\bar{1}1\}$ and $\{10\bar{1}2\}$ planes. Plasticity of an interface is therefore much influenced by the crystal structure.

17.1 *Adhesion Coefficient*

An adhesion concept has been used to show how the crystal structure of a material influences its friction and wear behaviour[1]. The phenomenon of metal asperities welding together can be described by an adhesion coefficient λ as

$$\lambda = \frac{\text{Force necessary to separate adhering solids}}{\text{Applied normal load}} \tag{17.1}$$

The adhesive bonding between materials is related to their modulus of elasticity and hardness, whereas the friction coefficient is dependent on the manner in which the interfaces will stick or slip. That is, if the sliding surfaces are favourably orientated so that the process is that of easy glide of the slip planes in the metallic crystal, sliding would be easy and a low coefficient of friction is expected.

17.2 *Experiment with Cobalt*

The role of crystal structure was investigated by using a hemispherical cobalt rider resting on a rotating disc of the same material[1] in a vacuum of 10^{-11} torr. The diameters of the rider and the disc were 1.3 and 6.35 cm respectively. Cobalt has a hexagonal structure and single crystals of this material were produced to eliminate the variables, the grain boundary. Sliding was such that the basal (0001) plane of the rider matched against a similar plane of the disc, the sliding velocity being 0.001 cm/s with a normal load of 50 g. The measured coefficient of friction was 0.35. To verify the probable role of the nature of deformation, the frictional characteristic of a fcc metal was studied[2]. In this, the disc and rider comprised single crystals of copper and the matching planes were {111} , which have the highest atomic density as the (0001) plane selected for cobalt. The coefficient of friction in this case was 21.0.

The reason for such divergent coefficients of friction under identical experimental conditions is given as follows. A fcc metal such as copper has 12 possible slip systems which increase the probability of gliding under shearing stresses. However, slip occurs by movement of dislocations and these, in intersecting slip planes, form a forest of dislocations which act as barriers to further slip. The idea thus is that, initially, asperities weld together readily because of easy slip. These junctions, however, are difficult to shear as they work harden because of the characteristic crystal structure and the frictional force is increased. Since the applied normal load is not altered, the coefficient of friction is high.

Such a physical model to explain the coefficient of friction seems feasible. Unfortunately, considering equation 4.13 it is difficult to reconcile this model with the fact that as the shear strength of the junctions increases their flow stress increases also. A more true picture is, probably, given by equation 5.8, since the work of adhesion in the fcc metals would be high giving a low value of the denominator in the equation. Hence a high coefficient of friction should be experienced.

With cobalt, a hexagonal metal, shear occurs between the basal planes. In this process, the dislocations do not pile up to form barriers to slip. Instead, they move readily out of the crystal. However, presence of solid impurities in the crystal structure is known to impede slip of these basal planes and the coefficient of friction then increases.

17.2.1 *Effect of Temperature.* Further evidence of the role of crystal structure in the friction behaviour of cobalt is given by continuing the experiments at high temperatures when, at a characteristic temperature, the coefficient of friction increases (Fig. 17.1). This is clearly due to the fact that cobalt transforms to the fcc structure at 417°C. It is interesting to note that the bulk temperature as measured by the thermocouple in the rider behind the sliding interface is more than 100°C lower than the temperature of the tips of the asperities which must be at or above the transformation temperature as evidenced by the rise in the coefficient of friction. This may be a useful tool to measure the temperature of sliding interfaces in terms of load and velocity.

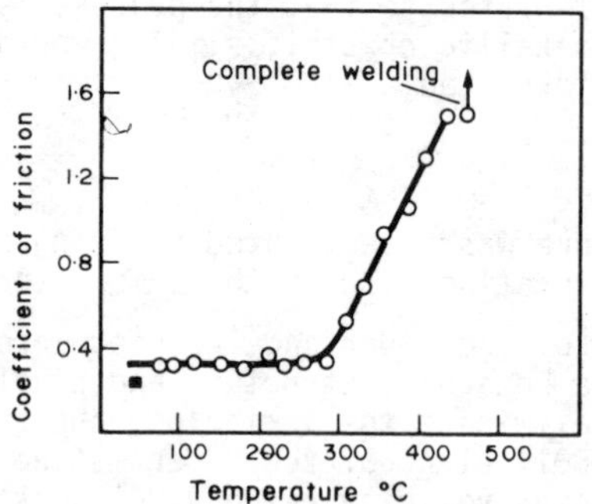

Fig. 17.1 Coefficient of friction for cobalt sliding on cobalt in a vacuum of 10^{-9} torr at various temperatures. Speed 198 cm/s. Load 1 kg (Buckley[1]).

17.3 *Rare Earth Materials*
A number of rare earth and related metals have been studied in a vacuum of 10^{-10} torr from the point of view of friction and wear[3]. It is again confirmed that the hexagonal structure gives lower wear and friction than the fcc

metals. The most favourable friction characteristics were obtained from those with the largest c - axis. The friction characteristic of sapphire sliding on itself is highly anisotropic, lower values being observed with basal orientations of the sliding surfaces[4]. Anisotropic behaviour in friction has also been observed with bcc tungsten[5].

17.4 *Change of Texture*

Materials in sliding contact are deformed and undergo a rise in temperature. This should cause recrystallisation in certain cases and give rise to a change in surface texture. Friction experiments[6] in a vacuum of 10^{-11} torr with both single and polycrystalline copper sliding on aluminium oxide at a speed of 0.001 cm/s showed recrystallisation of the worn track of both materials when the maximum load of 1 kg was applied. At lighter loads, x-ray analysis failed to detect any recrystallisation. However, lattice distortion in the subsurface was confirmed by electron diffraction. At heavy loads, the grains adjacent to the wear scars showed twinning, possibly as a result of annealing, since a fcc copper should not give rise to deformation twinning. Recrystallisation was observed in most metals at modest loads[7]. For example, tungsten, with a recrystallisation temperature of 1200°C, showed recrystallisation of the surface layer at a load of 3.5 g. No doubt future work will provide further contribution regarding the role of surface texture in the wear mode of metals.

REFERENCES

1. Buckley D H, *Cobalt*, (1968), *38*, 20.
2. Buckley D H, *Amer Soc for Testing and Materials*, (1968), Special Technical Publication No. 431.
3. Buckley D H and Johnson R L, *Trans ASLE*, (1965), *8*, 123.
4. Buckley D H, *Trans ASLE*, (1967), *10*, 134.
5. Buckley D H, *NASA Technical Note*, (1966), D-3238.
6. Buckley D H, *NASA Technical Note*, (1967), D-3794.
7. Buckley D H, *NASA Technical Note*, (1967), D-4143.

CHAPTER 18

ROLLING RESISTANCE

It has long been realised that the effort expended to transport objects by pushing them on rollers is much less than that by dragging them over flat surfaces. Use of rolling element bearing is, therefore, attractive because of the inherent low friction of these. Apart from bearings, examples of rolling situations are automobile tyres and railway wheels. This chapter outlines the two principal effects associated with rolling namely the slip or creep and energy loss, the latter being the measured frictional resistance.

18.1 *Principles of Rolling Motion*

Suppose that a hard sphere rests on a plane surface (Fig. 18.1) and, at the interface A, there is no friction. If a horizontal force, T, is now applied to the sphere, it will not roll because of the lack of interfacial friction. Instead, the sphere will slide in the direction of the applied external force. Similarly, if the sphere is rotated about its vertical axis, it will spin without sliding or rolling. As long as the contact at A is frictionless, sliding or spinning should continue indefinitely once started. As is the case in practice, there will be a finite amount of friction at A (Fig. 18.1) where the frictional resistance acts in a direction opposite to the external force T. This way, a couple will be established and the sphere will roll on the plane provided the applied force T does not exceed the interfacial frictional resistance F.

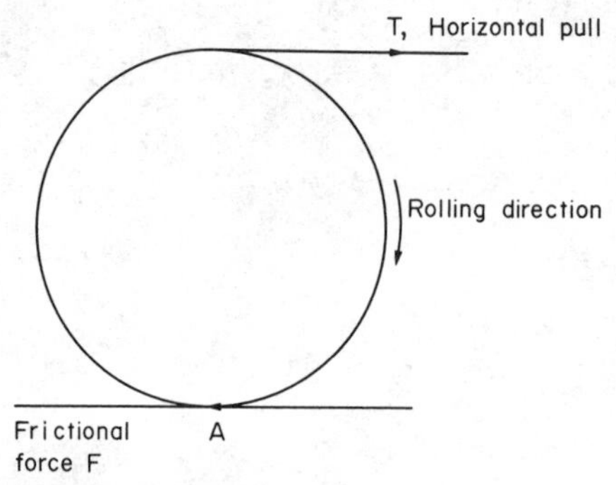

Fig. 18.1 A hard sphere resting on a plane surface.

18.2 *Slip*

Bodies in rolling contact under load and subjected to tangential traction show two regions in the contact zone [1, 2]. There is a part of the contact zone where both the sphere and the plane surface in Fig. 18.1 move together. That is, in this particular zone the two bodies are locked together and this fraction of the interface is termed the stick area. The remainder of the contact zone is the area of slip where there is relative movement between the ball and the plane. The former slides or skids in the direction of the tangential pull, or in other words, the plane surface moves backward in that part of the interface which is bounded by the area of slip. The reason for creep is attributed to the mismatching strain between the ball and the plane surface in the region where locking has occurred. Diagrams illustrating the boundary between these regions and the nature of strain pattern are shown in the next chapter.

As a general case, consider two cylinders rolling together. Let the normal load W indent a strip of area whose width is a. The width d of the strip over which the two cylinders remain locked together has been related[3] to T, the applied tangential pull per unit length as

$$d = a\left(1 - \frac{T}{\mu W}\right)^{\frac{1}{2}} \tag{18.1}$$

where μ is the static coefficient of friction between the two surfaces. Equation 18.1 shows that in the absence of a tangential pull, when T = 0, d = a, which means that the whole of the contact area is locked. If T equals the frictional resistance, d = o so that there is no locked region and slip takes place over the entire contact area. This is sliding or skidding on a large scale.

Apart from rolling with or without skidding at the contact area, a ball may spin around its axis. This is the case where the sphere and the plane have a relative angular velocity about this axis. Rolling, skidding and spinning are experienced in appropriate tribological situations.

18.2.1 *Reynolds' Slip.* The quantity $T/\mu W$ in equation 18.1 is really the ratio T/F where F is the static frictional resistance and the dependence of slip on a strain mismatch in the locked region is not apparent. Slipping due to a differential strain pattern is clearly demonstrated in what is known as Reynolds' slip. Reynolds[4] examined the case of a hard cylinder rolling on an elastic surface such as rubber (Fig. 18.2). Upon the application of load the cylinder indents a groove in the rubber and it is clear that extension due to compression is more in position 1 than that in the regions marked 2 and 3. If now a tangential pull is applied in the direction of the arrow the interface between A and B will slip because of this differential extension.

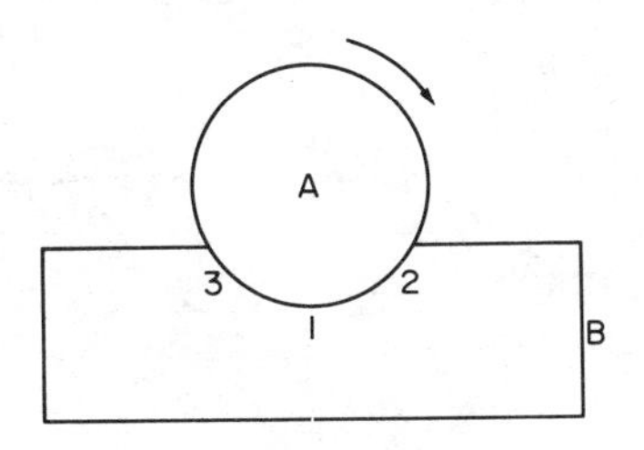

Fig. 18.2 A hard cylinder rolling on an elastic medium.

18.2.2 *Heathcote Slip.* Heathcote slip occurs for any materials when a sphere rolls in a groove (Fig. 18.3)[5]. The contact area is now elliptical and the relative velocity across the interface varies from point to point. This is readily appreciated by examining Fig. 18.3 since the outer part of the sphere rolls a lesser distance than the central region at any time interval. That being so, interfacial slip will occur. Geometric conformity between the ball and the groove requires an instantaneous axis of rotation and slip can occur in some region remote from this axis.

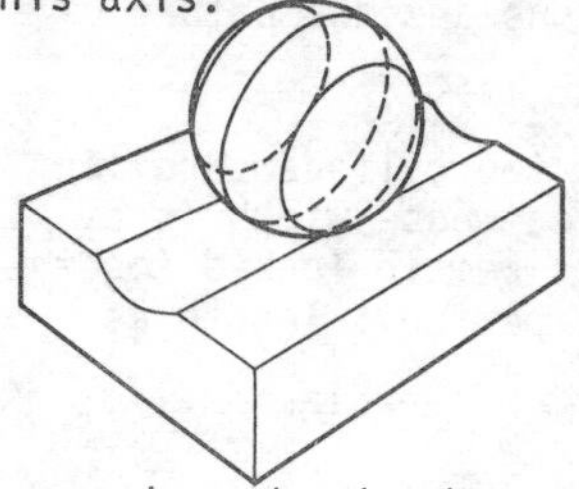

Fig. 18.3 Generation of a groove by a hard sphere rolling on a plastic medium.

18.3 *Rolling in the Plastic Range*

If a hard sphere is placed on a soft plate and then subjected to a normal load, the indented area will be a cup. The result of pulling the sphere will then be to generate a groove in the soft metal as is the case in Fig. 18.3. With repeated traversals, the groove will get larger as long as the soft metal is able to flow plastically. The groove will, however, work harden to a limiting value and the area will not change any further by plastic deformation since the interface is now elastic. A finite amount of force is necessary to overcome the interfacial resistance to rolling and this should depend on whether the system is in the plastic or in the elastic range.

18.3.1 *Track Width.* Rolling resistance has been studied experimentally[6] by rolling hard steel balls against various metals. As expected, the interface flowed plastically, being particularly marked for soft metals. A groove was generated by the ball in the soft surface after the first traversal and the measured frictional resistance was high. With continued traversals, the width of the track increased progressively (Fig. 18.4) but there was a corresponding fall in the resistance to rolling until an equilibrium state when both the track width and the frictional force became constant.

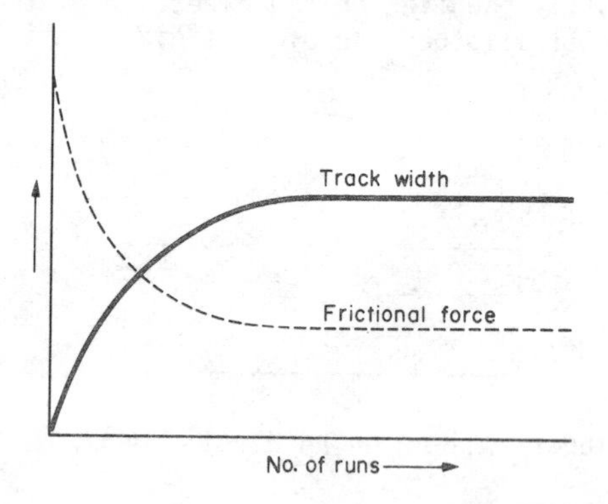

Fig. 18.4 Variation of track width and frictional force with the number of runs when a hard steel ball rolls against soft metals. (schematically drawn from the work of Eldredge and Tabor[6]).

18.3.2 *Rolling Friction.* There is thus an analogous situation to wear in that there is a running-in rolling resistance. Fig. 18.4 suggests that there is some connection between the track width and the friction at the interface. Since plastic deformation is involved, it is not unlikely that the applied normal load will influence the situation.

Consider Fig. 18.5. As rolling begins, the normal load W is supported by the area A_1, indented by the front half of the ball. If d is the track width,

$$A_1 = \frac{1}{2}\pi\frac{d^2}{4} \qquad (18.2)$$

The mean pressure acting on this groove assuming plastic flow is

$$p_m = \frac{W}{A_1}$$

and substituting for A_1 from equation 18.2,

$$p_m = \frac{8W}{\pi d^2} \qquad (18.3)$$

Note that the authors[6] do not bring in the concept of a true area of contact. That is, the interface is assumed to comprise truly smooth geometric solids, neglecting surface roughness.

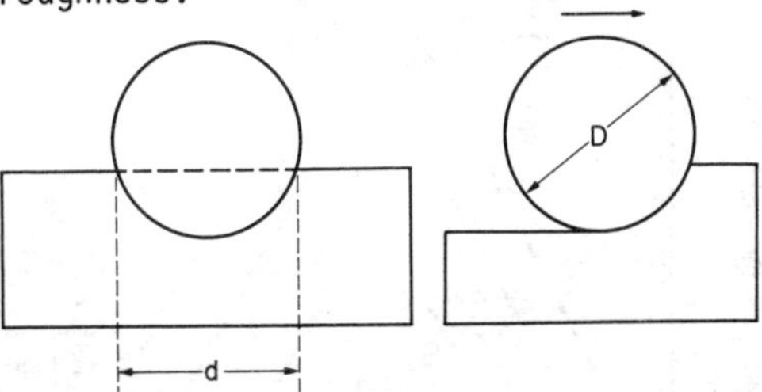

Fig. 18.5 Area indented by the front half of a rolling ball, diameter D.

As shown in chapter 4, the cross sectional area of the groove A_2 is $d^3/6D$ where D is the diameter of the ball. Making a further assumption that the tangential stress p_f resisting plastic displacement of the metal in front of the sphere is dictating the frictional resistance F,

$$p_f = F/A_2$$

or

$$F = p_f A_2 = \frac{p_f d^3}{6D} \qquad (18.4)$$

From equation 18.3, $d^3 = (8W/\pi p_m)^{\frac{3}{2}}$

Substituting this value of d^3 in equation 18.4,

$$F = \frac{p_f}{6D} \left(\frac{8W}{\pi p_m}\right)^{\frac{3}{2}} \tag{18.5a}$$

or

$$F = K \frac{W^{\frac{3}{2}}}{D} \tag{18.5b}$$

where K is a constant and

$$K = \frac{p_f}{6} \left(\frac{8}{\pi p_m}\right)^{\frac{3}{2}}$$, regarding p_f and p_m to be constants.

The difficulty with the foregoing is that it is known that part of the interface between the ball and the plane adheres. It is pertinent to ask if, unlike the analysis for sliding friction, the authors[6] are ignoring this adhesion term, probably because this is small compared to the ploughing term.

18.3.3 *Equilibrium State.* Equation 18.4 states that if the frictional resistance is plotted against $W^{\frac{3}{2}}$ or the reciprocal of the ball diameter, a linear relationship will obtain. This is so as is seen in Fig. 18.6.

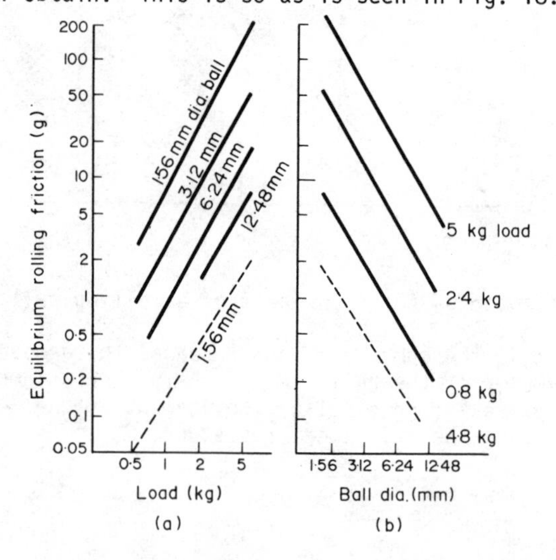

Fig. 18.6 Equilibrium rolling friction. (a) as a function of load at constant diameter of the ball; (b) as a function of the ball diameter at constant load. —— Annealed copper --- Hard steel (Bowden and Tabor, Friction and Lubrication of Solids, Part 2, Oxford).

In common with other modes of deformation, rolling gives rise to plastic flow initially but continued traversals create an elastic interface when there is no more plasticity. It has been suggested that metals which flow plastically reach an equilibrium state with time when subjected to rolling traction due to the following three factors:

(1) work hardening of the metal due to mechanical deformation;

(2) steady increase of the track width with time;

(3) a slight increase in the radius of curvature of the groove and this happens in the early stage when the metal is plastic; with most metals, the radius of curvature of the track is about 10% greater than that of the ball.

At the equilibrium state, the contact area is elliptical, the ratio of the axes being 5:1. This state, however, develops progressively and this is shown in Fig. 18.7. The final fully elliptical contact gives the equilibrium elastic deformation and the empirical law now is of the form

$$F = K \frac{W^n}{D^m} \tag{18.6}$$

where $n = 1.7 - 1.85$

$m = 1.5 - 1.70.$

Until the elastic or equilibrium state, the resistance to rolling is attributed to the effort expended for plastic displacement of the metal just ahead of the ball.

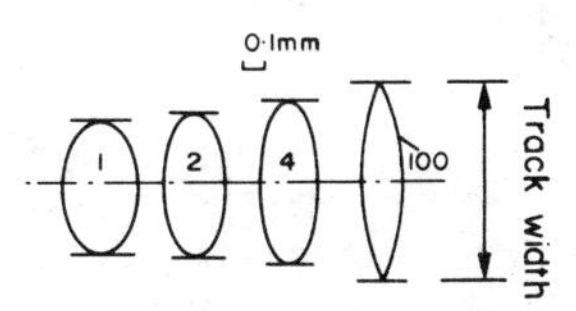

Fig. 18.7 Four stages in the development of the contact area generated by a ball rolling on a track. 1, 2, 4, and 100 traversals of the surface of annealed copper. Ball diameter 9.5 mm; Load 4 kg. (Bowden and Tabor, Friction and Lubrication of Solids, Part 2, Oxford).

18.4 *Rolling in the Elastic State*

Since the material has now work hardened, there is no more displacement of soft metal by the sphere and the ball rests on a curved track resulting in spinning and slipping in an elliptical contact area. There is still resistance to rolling and this aspect has been studied by a number of workers.

Tabor[7] has rolled cylinders and spheres on rubber to ensure an essentially elastic situation and also on hard metal surfaces. It is first argued that interfacial slip can not account for the rolling resistance because this is substantially the same for both lubricated and unlubricated surfaces. Furthermore, if a ball is rolled over another, there is no slip but the rolling resistance is the same as that obtained when one of the components is replaced by a grooved track.

18.4.1 *Hysteresis.* Tabor postulated that the measured frictional resistance encountered by a hard metal rolling on a similar elastic surface is due to hysteresis loss in the materials. To provide evidence supporting this view, a 0.312 mm diameter hard steel ball was rolled on a mild steel plate at a normal load of 4.8 kg. Plastic grooving finished in a few hundred traversals when the rolling resistance was 25 g with a track width of 0.45 mm, increasing further after 10,000 traversals. Increased track width of a groove should accentuate Heathcote slip and, if skidding and spinning were responsible, friction should rise. In fact, the value of the rolling resistance at that stage was 15 g and fell further to 12 g and 9 g respectively after 40,000 and 200,000 traversals.

If a metal is deformed in the plastic range and then subjected to cyclic loading and unloading, the hysteresis losses, high initially, are diminished gradually. The friction of the steel ball followed a similar gradual fall with time. Since experiments show that interfacial adhesion or slip makes negligible contribution to friction, rolling resistance at elastic contact is attributed to the hysteresis losses in the material during mechanical loading.

18.5 *Shake-Down-Limit*

If a cylinder is loaded against another only up to the elastic limit of the materials, the stresses everywhere in the body of the components will be elastic. If now the load is further increased, there will be plastic flow in the interacting solids. Since the cylinders are rolling, a point at the interface will become unloaded as soon as one of the components rolls out of the area. Because of plastic flow, however, an amount of residual stress will be left in the region loaded previously but now unloaded. The region will again be subjected to an external load during subsequent passes and the combined applied and residual stresses may not exceed the yield point of this region which is higher than the original material because of work hardening. That is, the system shakes down to a state of stress which is entirely elastic[8]. In a system such as the rolling of two cylinders, there is a maximum load beyond which the contact is non-elastic and this load has been called[8] the shake-down-limit.

When plastic flow occurs, it happens at a finite distance below the contact point so that a plastic zone is held within an elastic hinterland (Fig. 18.8a). Because of this situation, where an elastic outer layer is separated from a similar core by a plastic interface, a tangential pull should cause flow of material in the forward direction (Fig. 18.8b). Such a phenomenon has been observed experimentally by Crook[9], Welsh[10] and Hamilton[11]. Merwin and Johnson[8] have given a mathematical solution for a rigid cylinder rolling on a flat surface assuming that the deformation is plane and the material under stress is isotropic and elastic-perfectly plastic, that is non-work hardening.

18.5.1 *Forward Strain.* If the width of the contact zone is 2a, yielding of the material will first occur at a depth of 0.705a, when the maximum Hertzian normal stress $\sigma_{max} = 3.10\tau$, τ being the yield stress in simple shear or

$\tau = \sigma_y/\sqrt{3}$ where σ_y is the yield stress in tension. The shake-down-limit is shown to be reached when the maximum normal stress equals about 4τ. As this value is exceeded, the sublayer yields but there is no plastic movement in the direction normal to the contact area. Instead, the combined action of the residual and normal stresses produces an appreciable plastic shear in the direction parallel to the surface.

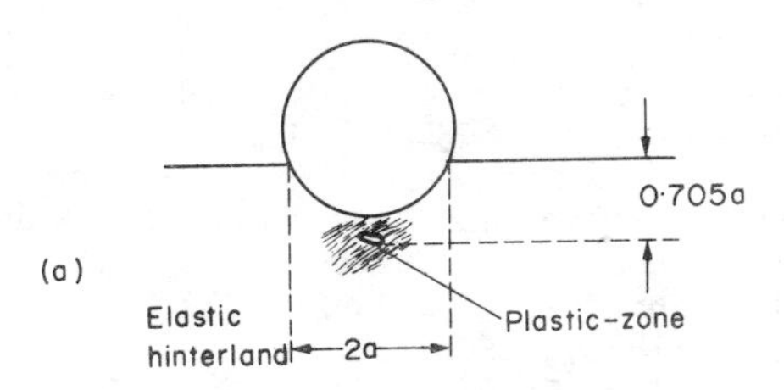

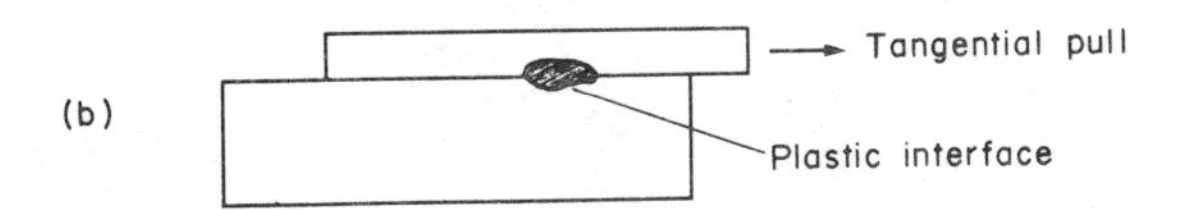

Fig. 18.8 A ball rolling on an isotropic and elastic-perfectly plastic medium. (a) A plastic zone surrounded by an elastic hinterland; (b) Flow of metal in the forward direction due to a tangential pull.

A physical model to describe the forward displacement of metal on the rolling element is given by the authors[8] as in Fig. 18.9, accepting that the cumulative deformation involved is by a simple shear process. The effect of the

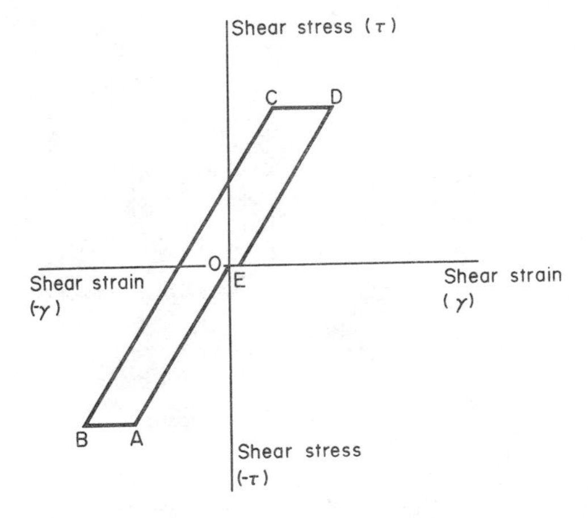

Fig. 18.9 Cumulative deformation of a rolling element by a shear process. (after Merwin and Johnson[8]).

rolling motion is that, during a loading cycle, the material will be subjected to an equal and opposite value of shear stress. Thus, at the leading edge, the contact zone is loaded up to the yield shear stress given by A ($-\tau$). Since there is plastic flow parallel to the surface, the result is a backward shear AB. As rolling continues, the stress reverses up to C, resulting in a forward shear CD. As the rolling element passes over, the stress reduces to zero at E leaving a permanent strain OE on the component in the direction of rolling.

REFERENCES

1. Giolmas S N and Halling J, *Proc Inst Mech Engrs,* (1964-65), *179,*145.
2. Haines D J, *Proc Inst Mech Engrs,* (1964-65), *179,* 154.
3. Johnson K L, *Jl Appl Mechanics,* (1958), *25,* 339.
4. Reynolds O, *Phil Trans,* (1876), *166,* 155.
5. Heathcote H L, *Proc Inst Auto Engrs,* (1921), *15,* 569.
6. Eldredge K R and Tabor D, *Proc Roy Soc A,* (1955), *229,*181.
7. Tabor D, *Proc Roy Soc A,* (1955), *229,* 198.
8. Merwin J E and Johnson K L, *Proc Inst Mech Engrs,* (1963), *177,* 676.
9. Crook A W, *Proc Inst Mech Engrs,* (1957), *171,* 187.
10. Welsh N C, *Proc Conf Lub and Wear, Inst Mech Engrs,*(1957),p.701.
11. Hamilton G M, *Inst Mech Engrs,* (1963), *177,* 1.

CHAPTER 19

WEAR UNDER ROLLING CONTACT

It is known that wheels of railways undergo wear mostly during acceleration, climbing a steep track and braking. Similarly automobile tyres wear during braking and quite severely during cornering. In the case of rolling element bearings the contact interface is deformed, the degree of which is determined by the material property, normal load and the amount of osculation, i.e., surface conformity. It does not follow that the surfaces of both elements conform faithfully and, consequently, interfacial slip occurs. This is likened to Reynolds' slip but, as discussed in chapter 18, slip occurs only over a part of the contact zone, the extent of which should be governed by the applied tangential pull and the frictional resistance at the interface (equation 18.1). Spinning and twisting can also occur due to design effect, e.g. a ball in a groove. Wear as a result of slip is anticipated in lubrication considerations but it has been pointed[1] out that wear resulting from this microslip occurs at high local stresses and very small slip velocities.

19.1 *Slip Area*

Wear occurs in the slip areas but both the normal pressure and the slip velocities vary over the contact zone. The normal stress varies because the contact is largely elastic and a similar pattern is expected for the shear forces across the contact zone. Thus consider Fig. 19.1 which shows[2] two cylinders producing a contact zone of width a under a normal load W. The area of no-slip is shown to be of width a′ located at the leading edge of the contact zone. The tangential surface force is T but the distribution of the tangential traction over the contact zone varies as shown. It is zero at the edge of the contact zone but rises to a peak at the boundary between the slipped area and that where the surfaces stick. The strain patterns of the two bodies have opposite signs because the tangential forces are in opposite directions between the rollers. Distribution of stick and slip regions has also been deduced[3] when a ball spins on a track about an axis.

19.2 *Wear*

Interfacial slip results in surface damage in the initial stage of rolling of a ball on a plane surface. Tabor[4] observed that, in his experiments, the interface did not show any obvious damage up to 100,000 traversals after which fine wear particles were produced. The amount continued to increase and the surface became rough with pits, possibly by fatigue after 200,000 traversals.

Wear mode of hard steel balls running both on harder steel and mild steel track has been studied by radioactive tracer technique[5]. A useful piece of observation was that, in order to get reproducibility of results, it was necessary to activate the whole body of the sphere rather than use mere surface activation. The initial loss of metal from the ball was by depositing material on to the track but back transfer of metal to the ball also occurred. There are suggestions from these experiments that wear debris probably formed from the deposit largely and the loss of metal from the ball was linear with the sliding distance. The mechanism thus would appear to be generally similar to that observed with brass on steel under sliding conditions. That

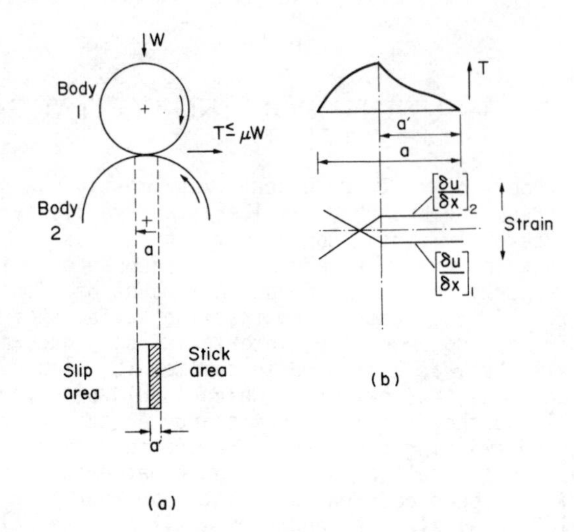

Fig. 19.1 Two cylinders rotating in the directions shown under a normal load W. (a) shows the contact zone; (b) shows the variable tangential force in the contact zone. The strain patterns for body 1 $\left(\frac{\delta u}{\delta x}\right)_1$ and of 2 $\left(\frac{\delta u}{\delta x}\right)_2$ in the contact zone are shown. u is surface velocity and x is a linear co-ordinate (Halling[2]).

is, the first effect is for metal to be transferred onto the track. Wear particles then detach from the deposit and an equilibrium state is reached when the rate of metal transfer equals the rate of debris production. However, the authors point out that the contact area for a ball rolling on a plane is the width of the track multiplied by the circumference of the ball. This is much larger than the track produced if the ball was made to slide continuously on the track. The result of this difference is that an equilibrium state in the wear mode occurs much earlier in the case of sliding than when the ball is rolling. It is also useful to note that metal transferred from the ball on to the track as discrete particles whose concentrations were the greatest at the centre of the band where the normal stress is the maximum. In a separate paper[6], the authors have reported that surface finish has little influence on the rate of wear.

19.3 *A Law of Rolling Wear*

Consider Fig. 19.2 in which a body A is rolling over another body B[7]. Assume that the mode of metal removal is by adhesion although it should be recognised that abrasion by debris may occur as the running time increases. For the purpose of analysis the authors[7] assume that metal is removed from the ball A only and that the body B does not wear.

Let

M_A = the amount of material removed from A but attached to it by, say, the back transfer from B and assume that M_A lies within the contact zone;

M'_A = material attached to A by a similar mechanism to M_A but lying outside the contact zone,

M_B = material transferred from A to B and held entirely in the contact boundary;

M'_B = same as M_B but lying outside the contact zone;

M_o = a wear debris removed completely from the system;

M_c = total amount of material attached to B;

M_w = total amount of material lost from A.

Thus,

$$M_c = M_B + M'_B \qquad (19.1)$$

$$M_w = M_B + M'_B + M_o \qquad (19.2)$$

Consider the case when up to a critical distance ℓ_1 traversed, there is no escape of wear debris from the system. That is, any material lost from A by interfacial slip remains attached at the contact zone or outside it. Thus for any length $\ell < \ell_1$, $M_o = 0$. Therefore, from equations 19.1 and 19.2

$$M_c = M_w$$

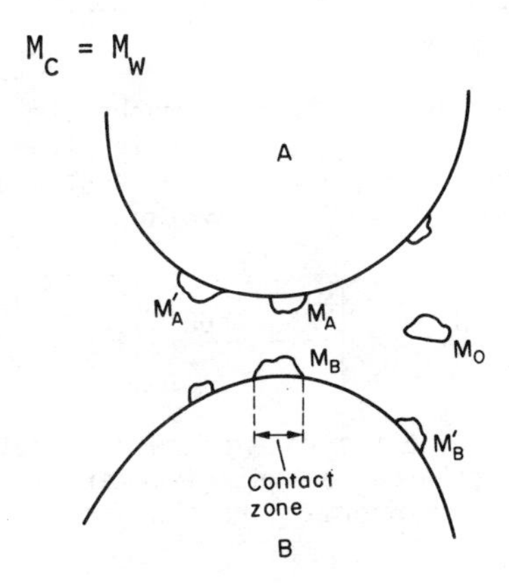

Fig. 19.2 A model showing wear of a cylinder, A, rolling on another, B (Giolmas and Halling[7]).

However, as the bodies in Fig. 19.2 continue to roll, an equilibrium stage in the surface interaction will be reached when the rate of increase in the metal transfer at the interface will equal the rate of escape of the product from the interface.

Now let

M_{AC} = the equivalent critical value of M_A and

M_{BC} = the equivalent critical value of M_B.

Then at $\ell \geq \ell_1$,

$M_A \equiv M_{AC}$ and

$M_B \equiv M_{BC}$

Assume now that the wear process at and beyond the distance ℓ_1 is given by

$$\frac{\delta(M'_A + M'_B + M_0)}{\delta\ell} = K\left[1 - \exp\left\{-\alpha\left(\frac{\ell - \ell_1}{\ell_1}\right)\right\}\right]\left(\frac{M_{AC} + M_{BC}}{\ell_1}\right) \quad (19.3)$$

K is a constant and α is regarded as the inverse of the transition time at a running distance ℓ_1. This time is very small so that, ignoring the exponential term, equation 19.3 can be written as

$$\frac{\delta(M'_A + M'_B + M_0)}{\delta\ell} = K\left(\frac{M_{AC} + M_{BC}}{\ell_1}\right) \quad (19.4)$$

K is a function of the nature of the wear mechanism and hence of the operating condition. Equation 19.4 shows a relationship between the rate of debris removed and that of material retained in the system and over the length interval ℓ_1 to ℓ, from equation 19.4,

$$M'_A + M'_B + M_0 = K\left(\frac{M_{AC} + M_{BC}}{\ell_1}\right)(\ell - \ell_1) \quad (19.5)$$

The above analysis shows that the rate of metal removed from the contact zone is such that a constant critical amount of material is maintained in the contact zone and the rate of debris formation is proportional to this critical amount.

REFERENCES

1. Halling J and Brothers B G, *Amer Soc Mech Engrs,* (1964), Paper No. 64 - Lub - 30.
2. Halling J, *J1 Mech Eng Sci,* (1964), *6,* 64.
3. Halling J, *Proc Inst Mech Engrs,* (1966-67), *181,* 1.
4. Tabor D, *Proc Roy Soc A,* (1955), *229,* 198.

5. Brothers B G and Halling J, *Brit J1 Appl Phys*, (1964), *15*, 1415.
6. Halling J and Brothers B G, *Wear*, (1966), *9*, 199.
7. Giolmas S N and Halling J, *Proc Inst Mech Engrs*, (1964-65), *179*, 145.

CHAPTER 20

POLYMERS

A popular material in bearing application is Polytetrafluoroethylene (PTFE) because of its low frictional properties. Bearings manufactured in PTFE alone are mechanically weak and hence, where load and speed specifications warrant it, the material is backed up with a metal liner. Porous metal bearings have also been impregnated with PTFE to provide low friction at the sliding interface. Apart from PTFE, polymeric materials find application in structural parts and they are used as synthetic fibres. The friction and wear characteristics of polymers are of interest and the salient features of their tribological properties are outlined here. The reader should look at the theories and experimental results in a comparative approach to those for metals presented in earlier chapters.

20.1 *Friction and Wear*

Experiments show[1] that most polymeric materials obey a relationship between the frictional resistance F and applied normal load W as follows,

$$F = \mu W^n \tag{20.1}$$

where n has a value of 0.80 for drawn nylon and 0.96 for cellulose acetate. The corresponding values for the coefficient of friction, μ , were found to be 0.92 and 0.60

Experiments[2] with PTFE sliding on itself have shown that the coefficient of friction increases with speed (Fig. 20.1). Figure 20.1 shows that the coefficient of friction does not vary with load but the values of μ are 0.02 and 0.32 respectively at sliding speeds 1.1 cm/s and 189 cm/s. These high frictional value were irreversible so that if the couples were later run at a low speed, μ still remained high and could only be lowered by removing the surface layers with a cutting tool before running the couple again. The authors[2] speculate that the phenomenon is probably tied up with frictional heating effect, resulting in viscous flow at the interface or a change of orientation of the polymer. The temperature effect is obvious in Fig. 20.2 where friction rises with temperature. It should be noted that μ is substantially constant up to a transition temperature when it rises sharply but assumes a steady value thereafter. The transition temperature band is narrow, being 15.5 - 17.5°C and 16.4 to 19.0°C for the low and the high speeds respectively.

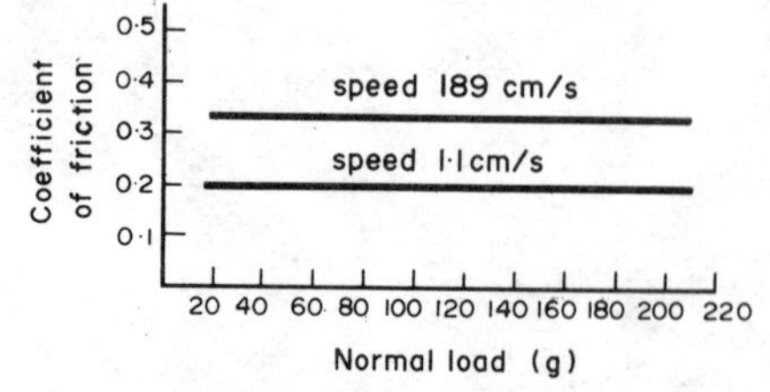

Fig. 20.1 Variation of friction with load for PTFE sliding on itself. (Flom and Porile[2])

There does not appear to be many published papers on the quantitative wear studies of polymeric materials. However, it has been reported[2] that, at a normal load of 108 g, the rate of wear was 4×10^{-11} cm^3/cm for PTFE sliding on itself and the wear rate was independent of pin geometry and sliding speed. An interesting observation is that if metal sliders are run on polymers, the former can be transferred on the latter if the sliding speed is high.

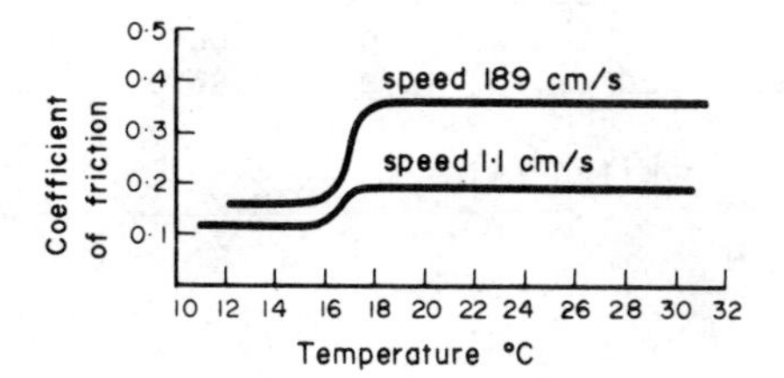

Fig. 20.2 Friction of PTFE sliding on itself as a function of temperature. (Flom and Porile[2]).

20.2 *A General Law of Friction*

It is known that the coefficient of friction for polymeric materials varies depending on the applied normal load. It is generally accepted that when the load is heavy, the interface flows plastically in which case the true area of contact is proportional to the applied load and the law of metallic friction is obeyed. As the load decreases, the contact becomes elastic with a consequent change in the mechanism of friction. These are the statements which are made to explain the fact that the coefficient of friction is constant at high loads but increases as the load is decreased (Fig. 20.3).

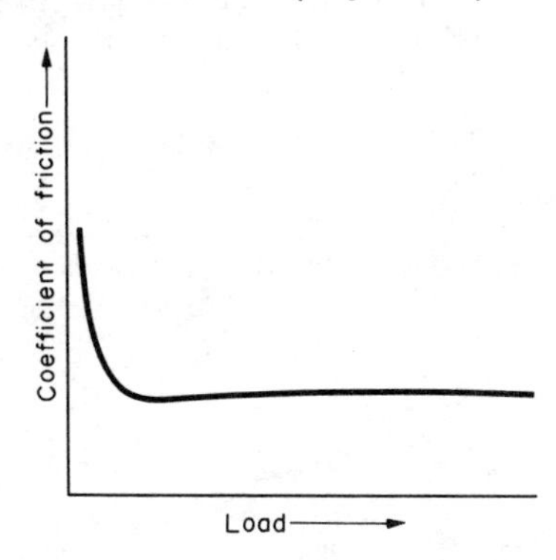

Fig. 20.3 Variation of the coefficient of friction μ with load. μ is constant at heavy loads but increases as the load is decreased (schematic).

Tabor[3] has studied the nature of local deformation on polymeric materials by using perspex in crossed cylinders configuration over a load range of 1 to 120 g. It was observed that the diameter d of the circular contact could be related to the normal load W as

$$W = Kd^m \tag{20.2}$$

where K is a constant and m has a value of 2.7.

As it is easier experimentally, in subsequent work, a steel sphere was indented into a flat perspex block. The load-diameter relationship was similar to that given in equation 20.2 except that, at any load, the indented diameter was about 20% greater than that obtained with perspex crossed cylinders.

If D is the diameter of the steel sphere, the geometrical description of the indentation can be given by the ratio d/D. That is, all indentations of the same d/D ratio are produced by the same mean stress σ and

$$\sigma = W/(\pi d^2/4)$$

That is,

$$\frac{W}{d^2} = f\ (d/D) \tag{20.3}$$

To solve the above equation, assume a solution of the following form:-

$$\frac{W}{d^2} = K \left(\frac{d}{D}\right)^x$$

where K and x are constants

or

$$W = K \frac{d^{2+x}}{D^x} \tag{20.4}$$

This, of course, is the general result for any material and is of the form given by equation 3.16. Putting m = 2 + x, equation 20.4 can be rewritten as

$$W = \frac{K\ d^m}{D^{m-2}} \tag{20.5}$$

Now the apparent area A of the circle of indentation is $A = \pi d^2/4$.

Substituting for d from equation 20.5,

$$A = \frac{\pi}{4} \left(\frac{1}{K}\right)^{\frac{2}{m}} W^{\frac{2}{m}} D^{\frac{(2m-4)}{m}} \tag{20.6}$$

The following assumptions are then made:

1. The frictional resistance is as a result of adhesion of asperities and the subsequent shearing of junctions.

2. The true area of contact A_t is the same as the area of the circle of indentation A.

3. There is no plastic displacement of junctions when the interface is

subjected to tangential traction.

4. The shear stress τ of the junction is a constant material property and is the same as that of the bulk polymer.

Now,

$$\mu = \tau A_t/W \tag{20.7}$$

Substituting for $A = A_t$, from equation 20.6,

$$\mu = \tau\left[0.78 \left(\frac{1}{K}\right)^{\frac{2}{m}} \left(W\right)^{\frac{2-m}{m}} \left(D\right)^{\frac{2m-4}{m}}\right]$$

or

$$\mu = 0.78\,\tau\left[\left(\frac{1}{K}\right)^{\frac{2}{m}} \left(W\right)^{-\left\{\frac{m-2}{m}\right\}} \left(D\right)^{2\left\{\frac{m-2}{m}\right\}}\right] \tag{20.8}$$

The numerical constant 0.78 in equation 20.8 is modified to 1.09 for the case where the components are crossed cylinders since the diameter is 20% greater. That is, the area has to be multiplied by $(1.20)^2 = 1.40$. Putting $\beta = (m-2)/m$, equation 20.8 can be expressed as

$$\mu = C\tau W^{-\beta}D^{2\beta} \tag{20.9}$$

This is the general law for friction of polymers at slow sliding speeds and the constant C depends on the macrogeometry at the interface.

An interesting explanation[4] for the load dependence of friction is that the interfacial shear strength of a couple increases with the external contact pressure. It has also been observed that the static area of indentation does not change in dimension with sliding. This is evidence to show that there is no junction growth in the case of polymeric materials constituting tribological components.

20.3 *Rubber*

The friction of rubber appears to be the same irrespective of whether the bodies are sliding or rolling if a lubricant is used to eliminate adhesion between surfaces, suggesting that the resistance to motion is entirely due to hysteresis loss of the material. It follows that vehicle tyres manufactured from materials with inherent high hysteresis losses will give better road safety, albeit with an increase in energy consumption.

The frictional resistance of rubber varies as the temperature and sliding velocity are altered. Various types of rubber were studied[5] under sliding conditions over the temperature range of -58 to +90°C. The rubber specimens were slid over rough silicon carbide papers and glass surfaces, prepared specially to give gentle wavy protuberances. In another series of tests, the silicon carbide paper was covered with a thin layer of magnesia powder to prevent any adhesion at the interface. The test piece was a pad of rubber, 25 mm square and 6.25 mm thick, the speed being restricted to a few cm/s to minimise frictional heating. The results showed the frictional resistance to be attributable to two factors, viz., adhesion at the interface and the hysteresis loss of the rubber. The temperature and velocity dependence of the friction of rubber was represented by a master curve as in Fig. 20.4 which shows a plot of the coefficient of friction μ against $\log a_T v$, where v is the sliding velocity and

$$\log a_T = \frac{-8.86(T - T_S)}{101.5 + (T - T_S)} \qquad (20.10)$$

and $T_S = T_g + 50$

where T_g is the glass transition temperature and T is the temperature of the experiment.

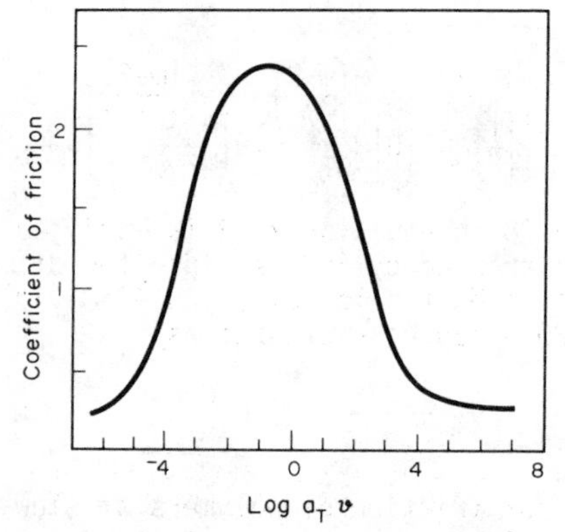

Fig. 20.4 Coefficient of friction of acrylonitrile-butadiene rubber compound on wavy glass. T_o = 20°C (see section 20.4) (Grosch[5]).

Figure 20.4 shows a maximum coefficient of friction followed by a fall. At low velocities or small values of $\log a_T$, the process is that of simple sliding but stick-slip motion is observed as the maximum is approached.

Slipping of wheels such as those on automobiles occurs frequently. The result of this is wear which is accentuated at high speed since the amount of slip increases at high tangential velocities. However, wear of car tyres can be reduced if a material with a high hysteresis loss is employed. A high hysteresis loss means a rise in frictional resistance. The corollary then is that the higher the frictional resistance, the lower is the rate of wear for rubber. It should be remembered that there is no obvious correlation between friction and wear in metals generally.

REFERENCES

1. Howell H G and Mazur J, *Jl Textile Institute*, (1953), *44*, 159.
2. Flom D G and Porile N T, *Jl Appl Physics*, (1955), *26*, 1088.
3. Tabor D, *Wear*, (1957-58), *1*, 5.

4. Adams N, *Jl Applied Polymer Science*, (1963), *7*, 2105.
5. Grosch K A, *Proc Roy Soc A*, (1963), *274*, 21.
6. Schallamach A and Turner D M, *Wear*, (1960), *3*, 1.

CHAPTER 21

FRETTING

In most mechanical, chemical or structural appliances, there may be surfaces which slip by a small amplitude relative to each other. This is not necessarily intentional and can happen due to vibration of machines resulting in an oscillatory movement, for example at a bolt joint. The result of this is fretting which is defined as a wear process occurring between two surfaces having oscillatory relative motion of small amplitude. There is now a body of knowledge on the aspects of friction, wear and lubrication for the situation where relative slip of small amplitude occurs between two surfaces and this has been presented in detail by Waterhouse[1]. In this chapter, a selected list of literature is reviewed to provide an understanding of the mechanism of fretting.

21.1 *Four Stages of Fretting*

As in experiments for studies on sliding wear, the magnitude of fretting has been evaluated by plotting weight loss of a sample with the number of oscillations. In most reported experiments, the amplitude of slip is varied since it has a decided influence on the degree of fretting, the upper limit being of the order of 0.2 mm. It is necessary to standardise the experimental procedure and the first requirement is to degrease specimens thoroughly before starting a fretting run. The samples can be weighed at suitable intervals, but in one set of experiments[2] with steel, the specimens were pickled in 5% sulphuric acid in water at 50°C containing an inhibitor quinolinethiodide. They were then scrubbed with a brush under running water, dried in acetone and then weighed.

The object of such a detailed cleaning process each time a sample was weighed was to remove any oxide film which formed during the rubbing of the surfaces. The pickling solution attacked the metal as well but a correction was applied to the weight loss to account for this. These experiments[2] established that the intensity of fretting was not dependent on the prior surface finish of the samples but the higher the relative slip, the greater was the surface damage. The effect of increasing the frequency of slip was to cause a rise in surface temperature and there was a decrease in the amount of wear when the experiments were conducted in dry air. A nitrogen atmosphere, however, showed no change in surface damage with the frequency of slip.

Plastic deformation, at least at the early stages of running, would appear to be an inevitable feature of all interacting surfaces and this has been observed[3] even on a raceway at the points of contact with a roller element. The first stage of wear has thus been attributed to metal transfer from one surface to another due to adhesion of asperities while examining the fretting process of mild steel on itself[4]. This is shown in Fig. 21.1 as OA. As wear particles detach and become comminuted, the mechanism changes to abrasion when work hardened debris remove metal from the surfaces. This is a transition range, AB. Beyond B, the rate of abrasive wear decreases, possibly because the surfaces of the steel specimens work harden also, giving the third stage of wear, BC. CD shows the steady state where the debris is generated at a constant rate.

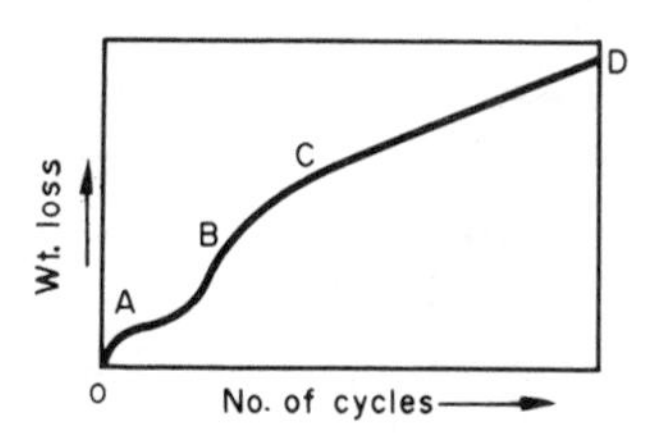

Fig. 21.1 Fretting as a function of the number of cycles. (Feng and Rightmire[4]).

The sequence of events during the relative slip process is shown schematically in Fig. 21.2 and the mode of wear is as follows:

(a) The applied normal load causes adhesion of asperities resulting in a situation as in Fig. 21.2a. As the contact areas slip, wear debris are produced which accumulate in the adjoining valley.

(b) The adhesive mode of wear gives way to the abrasive mechanism and the work hardened particles abrade the surrounding metal and the wear zone spreads laterally (Fig. 21.2b).

(c) When abraded sufficiently, the particles can no longer be contained in the initial zone and they escape into the adjacent valleys.

(d) The nature of the contact is now elastic because of work hardening and the maximum stress is at the centre so that the geometry now becomes curved as shown in Fig. 21.2d. This is a micropit and similar cavities develop in the adjacent valleys. As oscillation continues, these micropits coalesce to form larger, deeper pits.

21.1.1 *Worm Tracks*. Pits, roughly semicircular in shape, formed under oscillatory motion have also been referred[5] to as worm tracks. Experiments with mating steel surfaces with an initial Brinell hardness number of about 200 showed that, at small amplitudes of slip, there was no discolouring of surfaces which characterises fretting. The interface was damaged as would be expected in most sliding or rolling situations in the plastic range. At larger displacements, however, large fissures or worm tracks appeared in both surfaces. The fissures were roughly semicircular in cross section and extended across the surface, being perpendicular to the direction of movement. The debris showed discolouration as soon as the worm tracks appeared. The authors[5] were able to express the weight loss m with the radial displacement x by a relationship

$$m = ax + b\left\{1 - \exp(cx)\right\} \tag{21.1}$$

where a, b and c are constants. However, the weight loss values were not reproducible once worm tracking occurred.

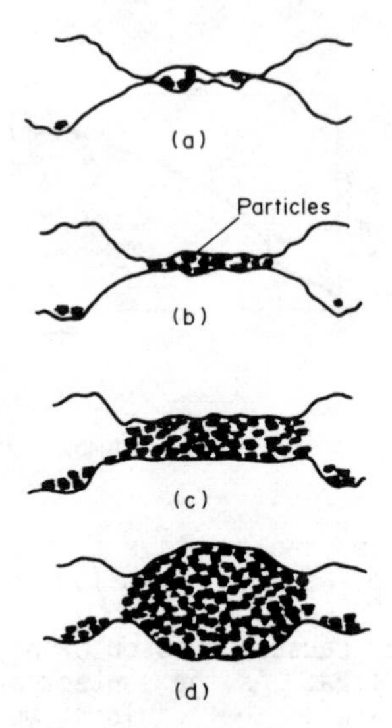

Fig. 21.2 Schematic representation of the initiation of fretting and the formation of large and deep pits. (Feng and Rightmire[4]). (a) trapped particles at the interface; (b) a number of contact areas wear to form a single large area; (c) spilling of particles in the adjoining depressed regions; (d) curved shape of a large pit formed by abrasion.

21.2 *Measurement of Pit Depth*

The colour of the detritus often indicates whether fretting has occurred. Thus the corrosion product of aluminium is white but fretting gives rise to black debris and, on steel, the product is red, being deeper in colour than that of rust. The pits produced during oscillatory movement of surfaces are sources of stress concentration and hence of failure of dynamically loaded components. Therefore, an attempt has been made to express the volume of wear by measuring both the wear scar and the depth of damage below it.

Wright[6] has done this by first measuring the damaged area and then establishing the loss of thickness and hence the depth of a pit by lapping the surface flat after a run. Wayson[7] has measured the depth of damage by carefully recording talysurf traces of surfaces which have undergone fretting. A typical trace is shown in Fig. 21.3 which also shows the heights of the metal fragments transferred from the opposing surface. Thus, while expressing the degree of fretting, a positive and a negative value is reported.

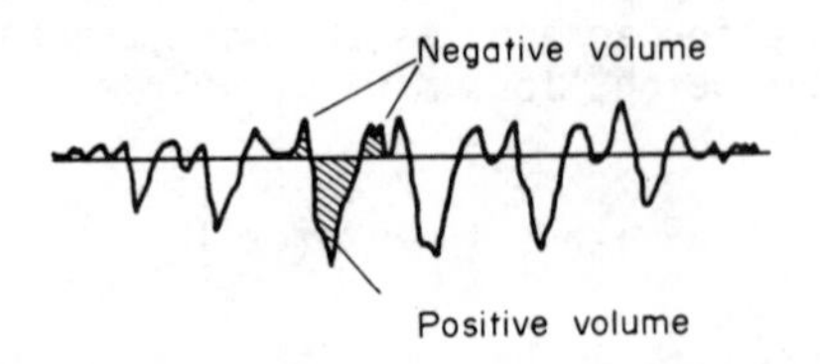

Fig. 21.3 Positive and negative volume from surface traces of a worn element (Wayson[7]).

A profilometric study such as this leads the author[7] to suggest three types of wear (Fig. 21.4) by measuring the fretting of platens of a number of metals against hard steel balls.

Type A: There is very little metal transfer and wear is even.

Type B: This shows heavy transfer of metal to the ball and deep platen wear.

Type C: This type shows wear but also a heavy build-up of metal. It is possible that the situation arises as a result of back transfer from the ball or by plastic flow of the platen surface.

A negative wear volume means deposition of metal and this is seen in type C wear. However, a negative wear volume is recorded in all types of fretting, especially in the early stages of running. As oscillatory movement continues, the transferred metal fragments disappear giving way to surface damage and pits only. It appears that the nature of surface damage is influenced by the platen material. For example, type A and B were typical of stainless and low carbon steels respectively.

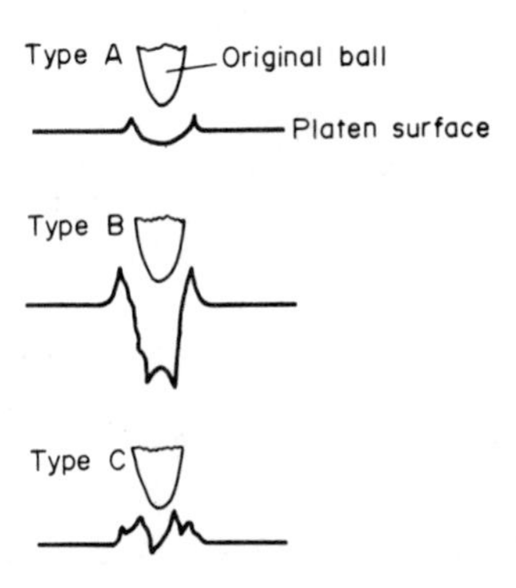

Fig. 21.4 Typical profiles of three major wear types (Wayson[7]).

21.3 *Load and Temperature*

The product of the area of wear scar and depth of pit has been shown[6] to vary with load in the range studied, viz., 10 - 60 kg at a slip amplitude of 2.5 x 10^{-3} cm for mild steel combination (Fig. 21.5). Using the same method to express wear, the amount of material removed was seen[8] to increase with temperature up to 400°C if the experiments were conducted in an argon atmosphere but the opposite effect was noted when the gas was replaced by air.

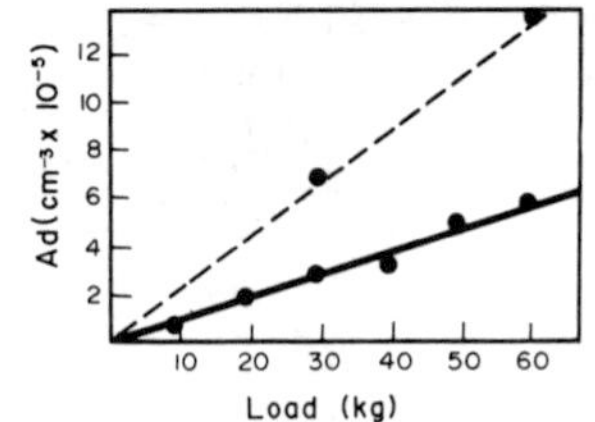

Fig. 21.5 Variation of the product of maximum area A and depth of fretting d with load. Steel on steel ------ 100% humidity; ——— 50% humidity(Wright[6]).

Light metals which have a low nuclear cross section for slow neutron absorption are of interest because compounds of unenriched uranium can be used. SAP, sintered aluminium powder, therefore, finds application in reactor technology. A part of the reactor system is so designed that the inner walls of SAP tubes are in contact with steel flanges and the coolant fluid, a mixture of terphenyls, flows through it at a high velocity. This causes vibration so that the SAP and steel are in a potential fretting situation at a high temperature. Simulated experiments[9] at temperatures up to 400°C have established metal transfer between SAP and materials such as steel, zirconium and stainless steel (Fig. 21.6). The degree of wear, however, could be minimised if the SAP surface was plasma sprayed with a 300 μ thick layer of Sb - Te alloy.

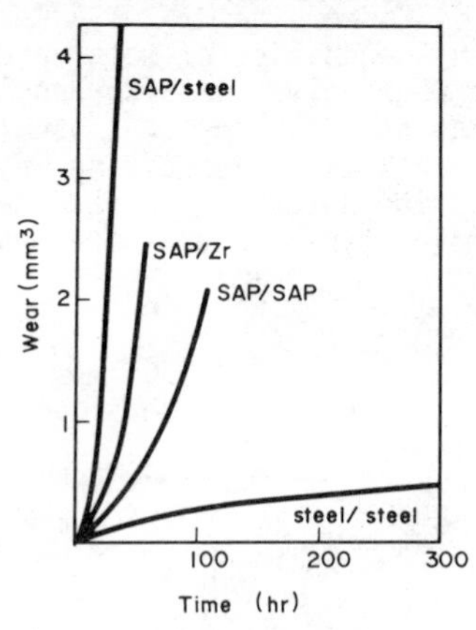

Fig. 21.6 Fretting with time of SAP against steel, SAP against zirconium, SAP against SAP and steel against steel (Commissaris and De Gee[9]).

In an extended programme[10], two types of motion were applied while rubbing SAP on itself:

(1) torsional vibration in the plane of contact; a fretting situation;

(2) vibration normal to the plane of contact.

The volume of metal removed was calculated by measuring the scar on a hemispherical rider with a radius of 10 mm which ran on a flat specimen. A terphenyl environment gave a somewhat greater wear than when the couple ran in a nitrogen atmosphere and the magnitude of fretting increased by a factor of 15 as the surrounding temperature was raised from 135 to 400°C. The mode of wear appeared to be adhesive in nature and the composition of the debris was the same as that of SAP. An appreciable increase in wear occurred if the debris produced was continuously removed from the interface. Figure 21.7 shows how the total volume of material removed varies linearly with the applied normal load. Vibration normal to the plane of contact gives a lower amount of wear than that encountered under fretting condition and the amount of material removed from the interface is the greatest when both types of motion are superimposed.

21.4 *Humidity*

Fretting is very sensitive to atmospheric humidity and this is reflected in the lack of reproducibility of results between experiments carried out in the

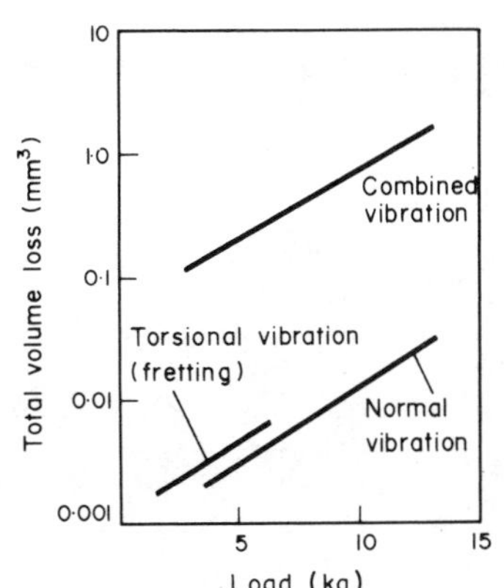

Fig. 21.7 Volume loss after 24 h as a function of normal load for various conditions of vibration. (De Gee, Commissaris and Zaat[10]).

summer and in the winter at room atmosphere. As Fig. 21.8 shows, surface damage decreases as the relative humidity of the atmosphere increases and it has been suggested that adsorbed moisture on the oxidised surfaces acts as a lubricant. Wright[6] suggests that a lubricating action may assist in removing the work hardened detritus from the interface, thus decreasing the abrasive component of wear. However, it has also been shown that removal of particles from the surface increases wear.

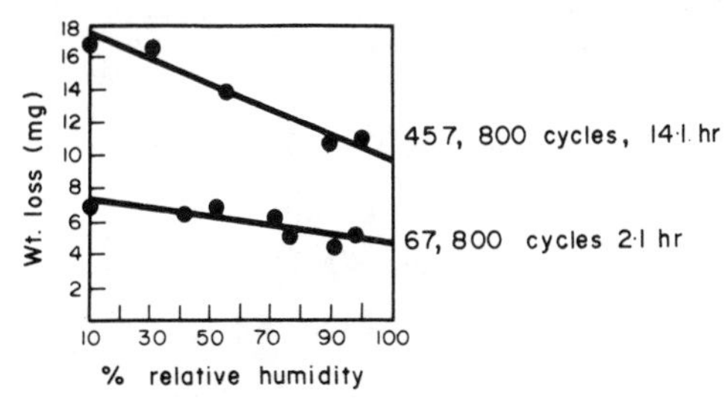

Fig. 21.8 Effect of humidity on the fretting of mild steel (Ming-Feng and Uhlig[2]).

Fretting can be minimised by interposing a non-corrosive lubricant at the interface and by surface treating the steel, e.g. phosphating. A PTFE washer inhibits fretting but polymethacryalate is ineffective. There is scope for research to obtain further insight into the mechanism of fretting.

REFERENCES

1. Waterhouse R B, *'Fretting Corrosion'*, Pergamon Press, (1972).
2. Ming Feng I and Uhlig H H, *Trans ASME - J1 Appl Mechanics*, (1954), *21*, 395.
3. Pittroff H, *Trans Amer Soc Mech Engrs*, (1965), *87*, 713.
4. Ming Feng I and Rightmire B G, *Proc Inst Mech Engre*, (1956), *170*, 1055.
5. Reed F E and Batter J F, *Trans Amer Soc Lubs Engrs*, (1960), *2*, 159.
6. Wright K H R, *Proc Inst Mech Engrs*, (1952-53), (1B), 556.
7. Wayson A R, *Wear*, (1964), *7*, 435.
8. Hurricks P L and Ashford K S, *Proc Inst Mech Engrs*, (1969-70), *184*, 165.

9. Commissaris C P L and De Gee A W J, *Proc Inst Mech Engrs*, (1966-67), *181*, 41.
10. De Gee A W J, Commissaris C P L and Zaat J H, *Wear*, (1964), *7*, 535.

CHAPTER 22

EXAMPLES OF TRIBOLOGICAL COMPONENTS

Apart from certain requirements such as load bearing capacity in common with most engineering structures, tribological components must meet the specified friction criteria. In certain situations a high value of friction must be met such as with braking surfaces, while in others a very important criterion is low friction as required for a shaft rotating in a bearing. Wear of interacting surfaces is beneficial during the running-in stage and can be profitable for a manufacturing organisation when the policy is to produce consumer articles or capital equipment with built-in-obsolescence. Unnecessary wear, however, must be minimised and the present and future society will, probably, demand components which will last and tribologists should aim at achieving a minimum loss in frictional resistance. To realise these objectives, the starting point should be to analyse the kinematics of interacting components in a tribological situation under examination. This, together with a knowledge of working load and environment, provides a sound basis for the choice of, inter alia, lubrication, material and manufacturing technique. In this chapter, a few components in common use are discussed briefly.

22.1 *Gears*

Gears are used to transmit motion from one part of a machine to another. There are various forms of gears but the principle of tooth-contact is described here using a pair of spur gears as an example. Figure 22.1 shows the successive stages of contact between two teeth, one from the driver and the other from the follower, superimposed on one diagram. Three positions are shown, viz., when the mating tooth profile is beginning a contact (A), when the contact is at the pitch point (P) and when this is just about to be broken (B). The path of contact APB begins at the tip of the driven gear and finishes at that of the driver. Gear teeth undergo both sliding and rolling.

22.1.1 *Sliding Velocity.* A point a on tooth A (Fig. 22.2) will have a velocity at right angles to Oa. The point of contact b on tooth B will have a velocity component at right angles to Qb. If ω_a and ω_b are the angular velocities of A and B respectively and ϑ_a and ϑ_b are the pitch line velocities of points a and b respectively,

$$\vartheta_a = \omega_a Oa \quad \text{and}$$

$$\vartheta_b = \omega_b Qb$$

The velocity of the pitch circle is

$$\vartheta = r\omega_a = R\omega_b,$$

where r and R are the pitch radii of A and B respectively.

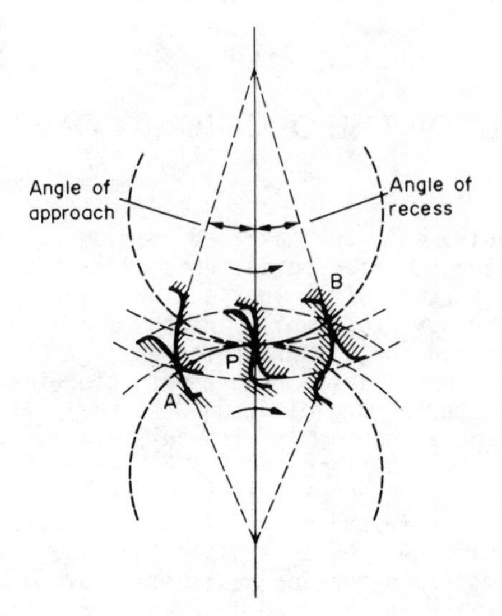

Fig. 22.1 Successive stages of contact between two teeth of the driver and the follower spur gear.

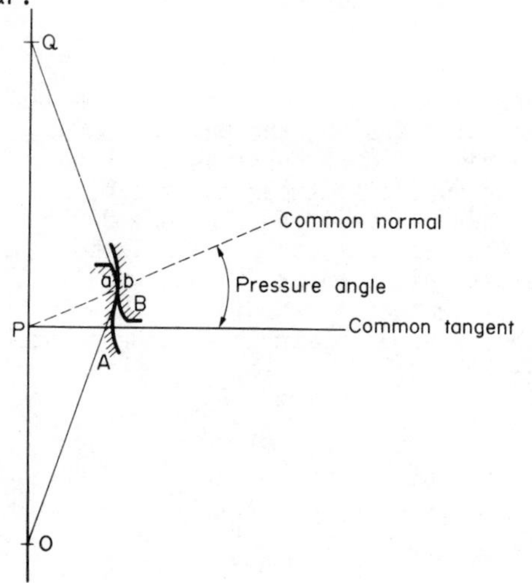

Fig. 22.2 Two mating gear teeth in contact.

The sliding velocity at the point of contact is the relative velocity of a and b and the direction is at right angles to Pa. If the distance between P and the point of contact at an instant is x, the sliding velocity ϑ_s between the teeth is given by

$$\vartheta_s = x(\omega_a + \omega_b) \tag{22.1}$$

The direction of sliding is reversed at the pitch point.

22.1.2 *Rolling Velocity.* A combined rolling and sliding action takes place when an involute spur gear moves relative to the other at the point of contact. This relative motion can be analysed by using two equivalent cylinders having surface speeds of v_1 and v_2 which are equal to the respective rolling velocities of the two teeth (Fig. 22.3) at that instant. As the cylinders are set in motion, the points a_1 and b_1 will arrive at the centre of contact 0 after a distinct time interval. The magnitudes of rolling for the two surfaces are given by the arcs a_1a_2 and b_1b_2 respectively and the amount of sliding is the difference between a_1a_2 and b_1b_2.

The sliding velocity is the difference between the velocities of rolling or sweep. As a tooth engages, the sliding velocity has a maximum value and decreases to zero at the pitch point. At this instant, the direction of sliding is reversed and the magnitude of the sliding velocity increases, reaching a maximum at the point of disengagement.

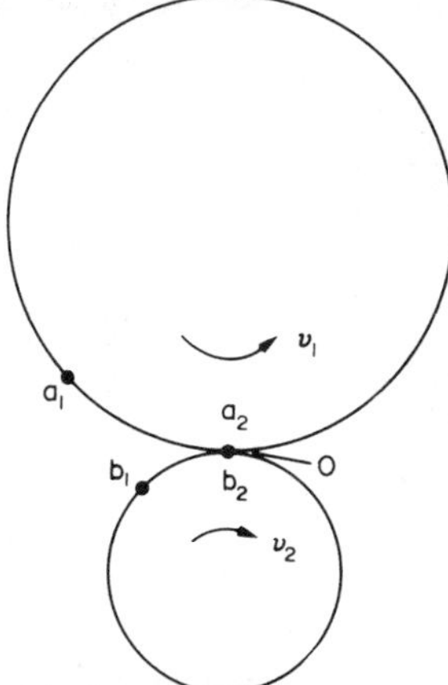

Fig. 22.3 Two equivalent cylinders to represent respective rolling and sliding velocities of two mating gear teeth.

22.1.3 *Friction and Wear.* As the analysis of the kinematics of gear teeth action shows, movement of gears involves both rolling and sliding at the flank and the face of a tooth. From first principles then, the total friction should be the sum of the rolling and the sliding component. However, both the direction of rolling and sliding and the corresponding frictional resistance vary during the interval when a gear tooth makes its first contact and later disengages. Measurement of the instantaneous coefficient of friction in relation to the meshing position has so far not been possible. Instead, an average value of the coefficient of friction, μ_a has been obtained empirically from results in disc-tests and used successfully in design work.

As the name implies, the test involves running two discs of radii r_1 and r_2 under tangential contact, usually with a lubricant at angular velocities of ω_1 and ω_2 respectively (Fig. 22.4). The circumferential velocities ω_1r_1 and ω_2r_2 can be regarded as the velocities of sweep at the point of contact between two meshing teeth, so that the ensuing sliding velocity v_s is

$$v_s = \omega_1r_1 - \omega_2r_2, \text{ assuming } \omega_1 > \omega_2.$$

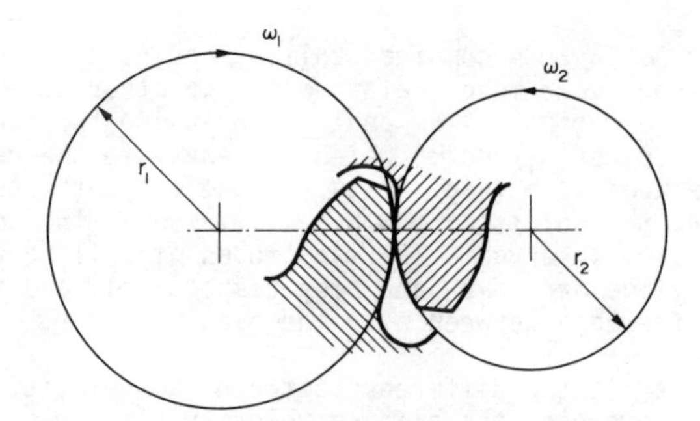

Fig. 22.4 Two discs to simulate the kinematics of gear tooth action (from Shell publications).

A relative radius r is then defined, such that

$$\frac{1}{r} = \frac{1}{r_1} + \frac{1}{r_2}$$

or $$r = \frac{r_1 r_2}{r_1 + r_2}$$

This can be compared with the relative radius of curvature, r_s, at the pitch line of a spur gear. Thus,

$$r_s = \frac{r_1 r_2}{(r_1 + r_2)} \sin \phi$$

where r_1 and r_2 are the respective pitch radius and ϕ is the pressure angle (Fig. 22.2).

From the disc test, the average coefficient of friction can be calculated from the following empirical formula:

$$\mu_a = \frac{K}{\eta^{\alpha}\, r^{\beta}\, v_e^{\gamma}} \qquad (22.2)$$

where v_e is the sum of the velocities of sweep, i.e. $v_e = \omega_1 r_1 + \omega_2 r_2$;
K is a constant, η is the viscosity of the lubricant, α, β and γ are constants with values less than unity.

Apart from the rolling and the sliding component, the sources of friction in industrial gears are resistance due to churning of the oil and windage loss. If lubrication is effective, sliding wear should not occur but trapped adventitious particles should result in abrasive wear, being aggravated by heavy load, high speed and temperature. A combination of heavy load and high speed gives rise to scuffing. Scuffing usually starts at the tips and the roots of the teeth and spreads towards the pitch line. The larger the tooth, the greater the propensity to scuffing since the relative sliding increases with the size of the profile.

22.1.4 *Pitting*. Pitting, often called surface fatigue, is very common in gears. The pits are generally conchoidal in appearance to the unaided eye and microscopic observations show further cavities in the vicinity of larger pits. Pitting normally starts in the neighbourhood of the pitch line and occurs during the running-in period of a gear. These early pits may be about 1 mm across and do not enlarge further once the surface stresses are uniform. These are referred to as arrested pits and, apart from these, progressive pitting may continue where the cavities increase in diameter and depth with time. The mechanism of pitting is attributed to the effect of alternate shear and tensile stresses at the points of rolling contact. The areas of damage are sources of stress concentration and, if progressive pitting occurs, layers of metal may peel off apart from the danger of breakage of the teeth themselves. It would appear that a good surface finish diminishes the tendency to pitting. There is some evidence to suggest that if the lubricant thickness is at least 3 times the cla value of the surfaces, pitting can be prevented provided the dynamic stresses are not high.

22.2 *Bearings*

The basic function of bearings is to locate a component accurately and support it under load without failure. Broadly speaking, a bearing may undergo a sliding action as in a plain bearing or a rolling motion as in rolling element bearings. It is not uncommon to run the bearings dry, but usually a lubricant is used to prevent metal to metal contact.

22.2.1 *Plain Bearings*. These can be sliding bearings, for example a machine tool slideway in which the work piece and the cutting tool can move relative to each other in a linear manner. In a plain journal bearing, the shaft rotates inside it and these can be vertical or horizontal. In the former, the nominal static load is zero and it can be very high in the latter.

Consider Fig. 22.5. The system has three components viz., the bearing, the shaft and a film of lubricant whose role is to partition the two. When the machine is not running, the lubricant has been squeezed out of the interface and the shaft rests on the bearing (Fig. 22.5a). As the machine starts, friction between the shaft and the bearing causes the former to lift (Fig. 22.5b). With continued motion, the applied torque and the increased thickness of the oil film overcomes metallic friction. At equilibrium (Fig. 22.5c), a hydrodynamic oil wedge is established and this has a pressure profile as shown in Fig. 22.5d.

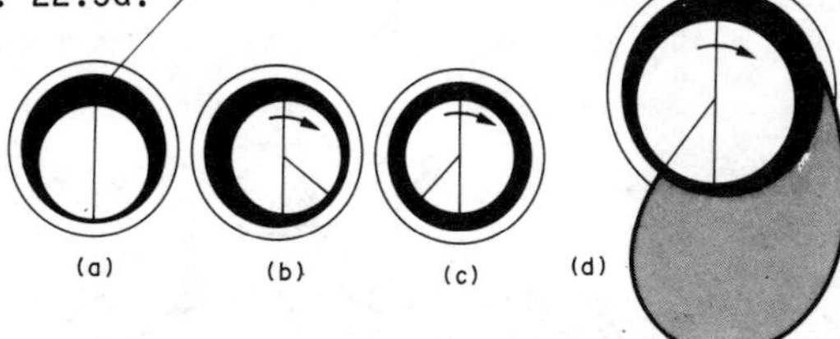

Fig. 22.5 Development of an oil wedge at the interface between the journal and a plain bearing. (a) machine at rest; (b) machine starts; (c) at equilibrium, formation of a hydrodynamical oil wedge; (d) the pressure profile of the oil wedge. (from Shell publications).

For a practical bearing, the coefficient of friction μ is related to the ratio $\eta N/W$ as in Fig. 22.6, where η is the viscosity of the oil, N the rpm and W is the applied normal load. At low speed or heavy load, lubrication is not so effective and the situation is that of boundary condition. An increase in speed decreases friction as the lubricant film builds up, reaching a minimum and μ rises again as the velocity or viscosity increases. This is a situation where the power consumption is high but metal to metal contact is prevented, diminishing the propensity to surface damage or wear. Boundary lubrication means that the possibility of metal to metal contact exists which will give rise to wear. This happens when an engine is first started or when it is idling and, incidentally, supports the need for research in dry sliding situation to draw definitive conclusions about metallurgical interactions in the kinematic chains of machines.

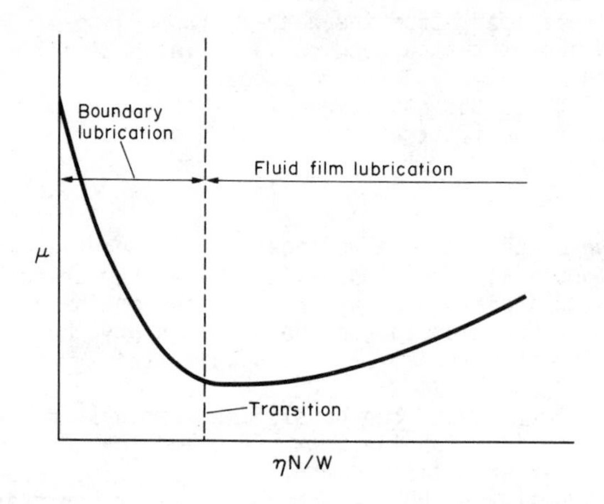

Fig. 22.6 Variation of the coefficient of friction between a journal and a lubricated bearing with speed, load and viscosity of the lubricating oil.

22.2.2 *Rolling Bearings*. Rolling, rather than sliding needs less effort and it has been stated that rolling bearings give the same order of low friction as encountered under fluid film lubrication in journal bearings. In a typical assembly (Fig. 22.7), the shaft is fixed to the inner race which rotates while an outer race is located in a housing. In the space between these races, the rolling elements are dispersed which may be in the form of balls or cylinders. The rolling elements are separated from one another by means of a cage which is also referred to as the separator.

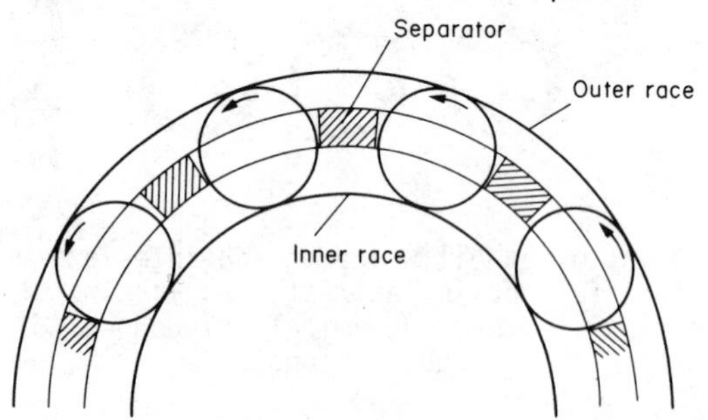

Fig. 22.7 Principles of a roller bearing (schematic) (from Shell publications).

As discussed before, skidding and spinning of the roller elements occur giving rise to wear which, however, can be minimised by the use of appropriate lubricants. A form of material loss from rolling elements is spalling of surface layers and is usually called fatigue, being aggravated by false brinelling which is indentation of surfaces arising out of vibration when the bearing is stationary.

22.3 *Piston Rings*
A typical reciprocating sliding situation is that of piston rings sliding on the cylinder liner. Carefully run flat piston rings develop a degree of ovality as shown in Fig. 22.8. This is as a result of wear and occurs due to a failure of lubrication. The nature of the profile of the lubricant film is shown in Fig. 22.9 which shows that the oil thickness becomes a minimum at the dead centres where the speed is approaching zero. The upper taper is assigned predominantly to temperature effect in the cylinder bore while in the vicinity of the top dead centre and the bottom taper is deemed to result because of the tilting of the piston in the bore as it traverses between the two dead centres. Formation of such a wear land is useful in holding an oil film.

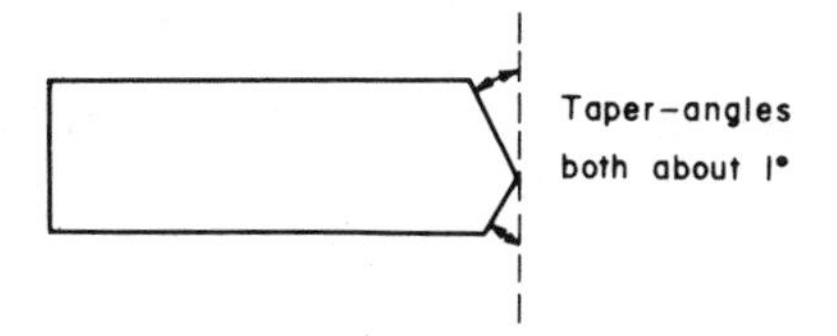

Fig. 22.8 The pattern of wear profile on a piston ring.

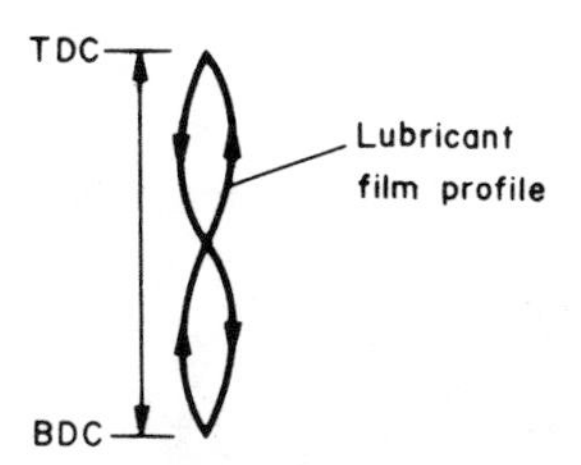

Fig. 22.9 Variation of the oil film thickness between the top and bottom dead centres of an I.C. engine.

If, however, the surfaces are not run-in carefully, severe damage to the interface can occur leading to scuffing. This is a serious surface damage and the only remedy is to ensure that the correct speed and load are maintained. An interesting observation is that a high wear rate may still occur in the absence of scuffing.

22.4 *Wear under Impact Condition*
In certain applications, for example jaw crusher plates, railway points and crossings, rock drills and excavator buckets, impact loading is expected but resistance to wear is very important if the components have to have an acceptable service life. Since it is shown that a material with a high flow stress wears less, a very hard steel could be recommended for the above com-

ponents. However, many alloyed hard steels lack ductility and the service requirement for a jaw crusher plate is that it should have a hard wear resistant surface layer with a tough inner core. This is ideally met in Hadfield manganese steel which contains about 1.2% carbon and 12% manganese. In the as cast condition, the steel has grain boundary carbides which introduce brittleness. Hence the cast components are made fully austenitic by heating them at 1000°C and quenching in water. Deformation in service such as abrasion by rocks gives rise to a layer of martensite in the surface which is hard and wear resistant but the core of the component remains tough. As the outer layers wear with time fresh martensite forms so that an adequate thickness of a hard layer is maintained continuously. A long life from these components is obtained if a martensitic layer is formed by explosive hardening before putting them in service.

The foregoing gives a few examples only of tribological components. The reader should consult Tribology Handbook[1] where many practical situations are cited if he needs knowledge about design and material philosophy of gears, cams, chains etc. in his career as a tribologist.

REFERENCE

1. Neale M J (Ed), *Tribology Handbook*, Butterworths, (1973).

CHAPTER 23

WEAR OF BRASS

60/40 brass is a good tribological material for experiments because it gives reproducible wear results. Brass has been used by many workers to study the mechanism of wear and it is a good material to use whenever the reliability of a new rig is being tested.

23.1 *Weight Loss With Sliding Distance*

A typical plot of loss in weight of brass pins running on steel bushes is shown in Fig. 23.1 which shows a running-in stage followed by a steady state wear. There are two aspects to note in experiments like these. Firstly, replicate measurements show that the magnitude of the running-in wear is not reproducible, suggesting that the law of Queener et al (equation 7.6) does not describe this regime completely. Secondly, the slope of the steady state wear is quite reproducible.

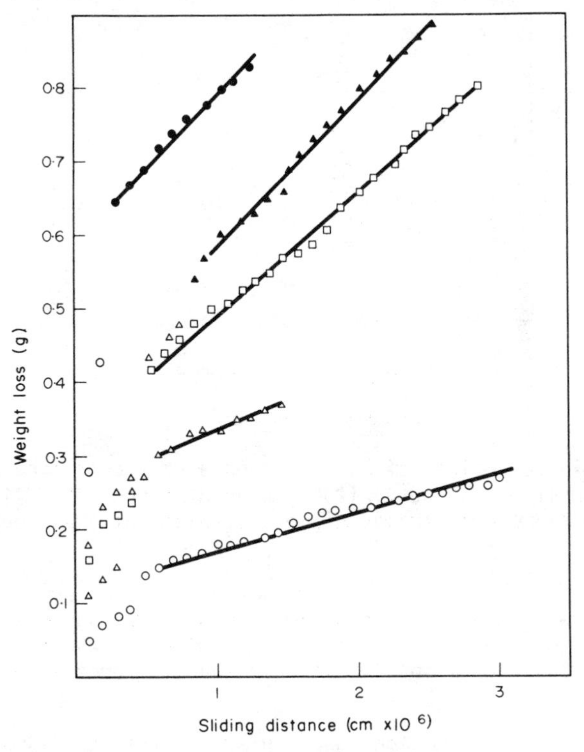

Fig. 23.1 Weight loss of brass pins with 45^0 cone angle sliding on mild steel bush. Surface speed 59 cm/s. 60/40 brass with 0.5% lead. Loads in kg O 2, △ 3; □ 4; ▲ 5; ● 6.

Pins of various geometry can be used. A useful configuration is to use a stationary crossed cylinder of brass and follow the change in diameters of the ellipsoidal scar at short time intervals. Typical results are shown in Fig. 23.2 which also shows the width of the deposited brass on the bush. The axes of the wear scar increased with time but, if measurements are taken frequently, it can be seen that there are instances when either the minor or the major axis ceases to change in dimension. It is permissible to speculate that material is removed by adhesion or abrasion followed by work hardening when wear ceases. However, wear commences again with continued sliding and this may happen because of either or both of the following phenomena.

(1) The work hardened layer is removed by a fatigue effect and fresh soft surfaces are exposed.

(2) The heat evolved by friction causes work softening which decreases the flow pressure of the junctions and wear proceeds according to equation 8.9.

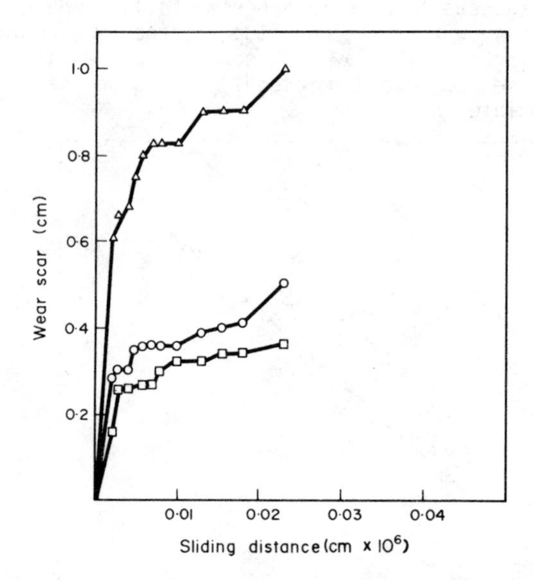

Fig. 23.2 Change in dimensions of wear scar on a crossed cylinder and the width of brass deposit on the bush with sliding distance. Load 14 kg; speed 85 cm/s. o minor diameter; Δ major diameter; □ width of brass deposit.

23.2 *Wear Rate*

The wear rate of brass as a function of load (Fig. 23.3) shows an independence of this parameter from the macrogeometry of the pins. The wear rate is near linear up to the transition load of 8 kg when the pin loses material at an accelerated rate.

Tabor[1] has shown that the Brinell hardness H of a metal is related to its yield pressure σ_y as,

$$H \simeq 3\sigma_y \qquad (23.1)$$

Therefore, the law of adhesive wear can be written,

$$\frac{V}{S} = \beta \frac{W}{H} \tag{23.2}$$

To preserve dimensional homogeneity, multiplying both sides by the density of the material under study, the rate of wear q in g/cm is

$$q = \beta \frac{W}{H} \rho \tag{23.3}$$

Taking the hardness of brass as 12.7×10^{-6} g/cm^2 and its density as 9 g/cm^3, the wear coefficient β has a value of the order of 10^{-4} up to a load of 8 kg, which compares fairly well with the experimental result of Archard[2].

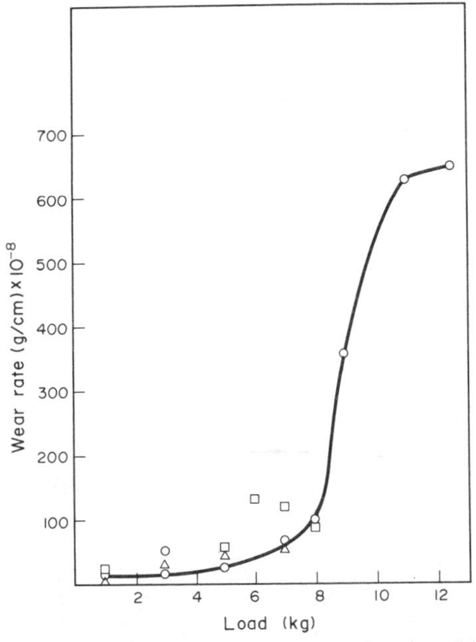

Fig. 23.3 Wear rate of brass pins at various loads sliding on hard alloy steel bush. 60/40 brass with 0.5% lead; speed 59 cm/s. o vertical cylinder; □ crossed cylinder; △ hemisphere.

The direct proportionality of q on W is based on the assumption that, in order for a wear fragment to be produced, the contact must be plastic or metal removal should be by a process of abrasion. Observations on brass sliding on steel demonstrate that metal removal from the pin is a process where cold welding is involved. This means plastic flow but as the pin slides on the brass deposit, stick-slip becomes accentuated and gross deformation of the subsurface of the pin occurs, especially if the load is heavy. A hard surface layer also forms which is capable of abrading mild steel but not the harder steel surfaces. The layer is extremely hard and brittle and, therefore, rather than flow plastically, it can only bend in an elastic manner or fail by brittle fracture. Since radioactive tracer studies show[3] that metal transfer takes place throughout a run, this probably happens in those parts of the pin where the brittle layer has peeled off. These exposed patches soon become covered with a hard layer but other areas are exposed for metal transfer to take place. Therefore, it would appear that a wear fragment is produced by both plastic and elastic effect and, in the latter case, it is possible that a wear debris will result when a critical amount of energy is stored by a domain which has exceeded its force of adhesion to the substrate.

It has been recognised by many workers that a wear debris may be produced by either plastic or elastic encounter and Archard[2] shows the probable relationships between the rate of wear q and the applied load W according to the nature of deformation of the sliding surface. This is summarised in table 23.1, which shows that a distinction is also made between the shapes of the particles, that is whether they are detached in lump or in layer form.

Table 23.1 Wear Rate q as a Function of Load W (Archard[2])

Deformation	Particle shape	q = f(W)
Elastic	Layer	$q \propto W^{0.60}$
	Lump	$q \propto W^{0.80}$
Plastic	Layer	$q \propto W^{0.75}$
	Lump	$q \propto W$

It is not easy to differentiate between the layer and the lump form of wear debris but, generally, flake like particles detach at low loads giving way to a coarse product when the apparent contact pressure is high. At low loads, therefore, a direct proportionality of the wear rate with load should not be expected.

23.3 *Transition Load*

Figure 23.3 shows that, at a load of about 8 kg, a marked transition occurs giving heavy wear. The diameter of the cylindrical pin loaded vertically in the experiments which produced the results for Fig. 23.3 was 0.625 cm so that the apparent contact pressure for the transition point is 30 x 10^3 g/cm^2. This is not 1/3rd the hardness of the brass which is the value where transition is shown to occur for steel sliding on steel by Burwell and Strang[4]. At a first glance, the observation of these authors[4] is very attractive since the flow pressure of a metal is 1/3rd its hardness and, therefore, gross adhesion must take place at the right contact load. It is true that if a metal component is loaded such that the applied stress exceeds its flow pressure, bulk deformation of metal will occur. This has important technological application in the field of friction welding. For a sliding situation, it may be called catastrophic wear but it is reasonable to ask whether it is fair to classify it as a wear process. If a situation such as this happens in machine parts, it is simply a case of failure by seizure and wear should be defined as a process of metal removal where the mechanical stability of the bulk material is little affected.

Considering Fig. 23.3 and taking into account the experimental results and theories of wear of various workers the probable wear rate - load characteristics are shown schematically in Fig. 23.4. OA is the rate of wear at moderate loads and is near-linear. This is mild wear because the junctions are diluted by oxide films or atmospheric gases but the fragments must detach from the substrate either or both according to the laws of adhesive and abrasive wear. As the load is increased to 8 kg, the contact pressure at the junctions becomes large enough for yielding to occur in depth giving wear fragments in lump form and, since contact is plastic, there will be a direct proportionality to load (table 23.1). It is interesting to note that the true contact area is some 10^{-2} to 10^{-3} times the apparent area[5] which in this case is 0.28 cm^2. Therefore, the true contact area for the pressure at a load of 8 kg to be equivalent to 1/3rd of the hardness of the brass must be about 2×10^{-3} cm^2.

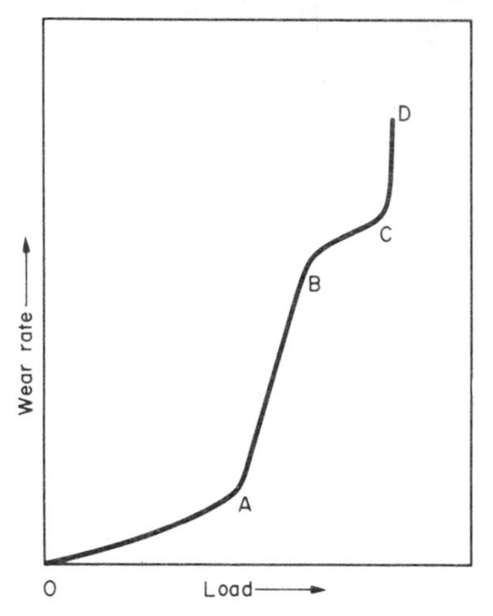

Fig. 23.4 Schematic representation of the probable variation of wear rates of brass against steel. OA, approximately linear; AB steep rise, a transition state, which is approximately linear; BC not necessarily linear; C → D, seizure due to bulk deformation.

Sliding interfaces become hard even at very small loads and metallography or microhardness measurements of transverse section show a hard subsurface. Increasing the load beyond B in Fig. 23.4 produces a hard phase with sufficient coherency for it to protect the underlying metal and wear particles are probably produced by a form of brittle failure and the rate of metal loss slows down. At a point C, the flow stress of the bulk material, as measured by tensile testing, is reached and severe cold welding on a macro scale occurs causing catastrophic damage and, possibly, seizure.

The hard layer formed on the brass appears dark and extends into the body of the pin. Lancaster[6] has identified a similar layer having a composition of 38 wt % iron. He also established that the zinc oxidised preferentially and the concentration of iron decreased with depth from the surface. The presence of a partitioning film such as this should give a diminished wear rate in the region BC in Fig. 23.4 but radioactive tracer studies[7, 8] show that the wear mode of brass is largely due to metallic contact meaning that at all stages welded junctions form, albeit with the diluting effect of oxides and contaminants. On the other hand, wear process in steel sliding on steel is shown to be governed by the removal of oxide films which form at the interface. This suggests that Burwell and Strang[4] observed mild wear up to the yield pressure of the steels at which point the bulk material became unstable (Fig. 8.1). Eyre and co-workers[9], however, show a transition regime at a much lower load than that of Burwell and Strang[4] while sliding steel and cast iron on steel.

REFERENCES

1. Tabor D, *'The Hardness of Metals'*, Oxford University Press (1951).
2. Archard J F, *Jl Appl Phys*, (1953), *24*, 981.
3. Kerridge M and Lancaster J K, *Proc Roy Soc A*, (1956), *236*, 250.
4. Burwell J T and Strang C D, *Jl Appl Phys*, (1952), *23*, 18.
5. Bowden F P and Tabor D, *Proc Roy Soc A*, (1939), *169*, 391.
6. Lancaster J K, *Proc Roy Soc A*, (1963), *273*, 466.
7. Archard J F and Hirst W, *Proc Roy Soc A*, (1956), *236*, 397.
8. Archard J F and Hirst W, *Proc Roy Soc A*, (1957), *238*, 515.
9. Hodgson M J and Eyre T S, *Jl BCIRA*, (1967), *15*, 257.

CHAPTER 24

FRICTION AND WEAR OF GRAPHITE AND CARBIDE

Graphite and carbide find application in tribological situations as such components as bearings and cutting tools. Graphite is presumed to offer low friction whereas carbides, being hard, provide the logical basis for applications where wear resistance is desired. Apart from this individual use, both graphite and carbide form important constituents in a technologically important material, cast iron. The available literature on the friction and wear mode of these two materials, therefore, is reviewed briefly in this chapter.

24.1 *Graphite*

Graphite possesses a hexagonal crystallographic structure and its good lubricating property in air was attributed to the easy shear of the loosely bound layer of this lamellar solid. However, graphite has a high interlamellar binding energy in vacuum and it is a poor lubricant under that condition[1] so that the material must undergo a weakening of its interlamellar bond strength in an environment of air or water to explain its good lubricating property in room atmosphere. One proposed mechanism[2] is that when two graphite surfaces approach each other, the charge mosaics on them effect a net attractive force but, as gases such as air are adsorbed on the surfaces, some of the charge is neutralised and the interlamellar binding energy is lowered. In high vacuum, the adsorption rate of gases is extremely slow and this explains graphite's high cleavage strength and poor lubricating properties in that environment.

Savage[3] has demonstrated the effect of adsorbed gases on the frictional behaviour of graphite by rubbing rod like specimens 1.8 to 6.4 mm^2 cross-sectional area against 15 cm diameter copper discs rotating with a peripheral speed of 1370 cm/s. The disc was cleaned chemically in a sodium carbonate solution and the graphite sample was freed from contaminants by baking it at 400^oC. The debris produced was in dust form indicating that fragmentation of the graphite crystals occurred. The coefficient of friction was high at μ = 0.8 when the experiment was carried out at low pressures but a reversible drop in the value to 0.18 was observed when water vapour was introduced into the atmosphere. The wear rate of 0.1 mm^2/s in vacuum fell to almost zero at a water vapour pressure of 3 mm mercury. Oxygen minimised wear similarly although the pressure necessary to give a comparable wear rate to water vapour had to be increased by a factor of 100. Generally, condensible vapours such as ammonia, acetone, benzene, ethanol and diethyl ether lowered the rate of wear. Hydrogen, nitrogen and carbon monoxide showed no lubrication effect.

The reason why certain gases are able to lubricate and thus minimise wear is explained[3] with the aid of condensation-evaporation theory. This states that the rate of condensation must be equal to or greater than the rate of evaporation in which event a transient monolayer of the gas would form on the graphite specimen. This would reduce the surface energy of the sliding surface and hence the coefficient of friction. The calculated value of this condensation life could be of the order of 10^{-6} second. It is suggested that a gas like hydrogen which has a very low boiling point possesses a condensation life too short to remain as a monolayer at the graphite-copper interface.

Whether the copper disc was regularly measured for surface roughness is not stated but it may be assumed that this was sensibly constant since all surfaces were prepared cathodically. The author[3] observed that the disc was covered with a layer of copper oxide and upon this was a thickness of graphite dust so that rubbing, in effect, was between graphite and graphite. The rather high value of the coefficient of friction at $\mu = 0.18$ in an atmosphere of water vapour, which gives little wear, is attributed to the force necessary to separate a coherent water film at the interface for sliding to occur. There was evidence to suggest that the frictional heat was conducted away from the copper disc because of its superior thermal conductivity but to a much lesser extent when a graphite rod was rubbed against a graphite disc. In the latter case, the coefficient of friction increased with surface speed showing an abrupt transition region.

In a separate experiment[4], it was demonstrated that the efficacy of condensible vapours as lubricants to graphite depended on the ratio p_1/p where p_1 is the vapour pressure of the gas at which the rate of wear is zero and p is the saturation pressure of the vapour over its liquid at the operating temperature.

It was noted that the lubrication efficiency of a condensible vapour increased with the chain length of the compounds. Thus, methanol with a chain length of 6.14Å requires a gas pressure of at least 0.65 mm Hg to be effective as a lubricant compared with n-heptane of chain length 13.88Å whose minimum lubricating pressure is 8×10^{-4} mm Hg. The latter pressure is equivalent to only a few parts per million of n-heptane in ordinary air. This means that when graphite slides on metals in room atmosphere, the presence of even trace amounts of the right type of contaminating fluid would affect the wear mode of this lamellar solid markedly.

The friction coefficient of outgassed specimens of graphite under reciprocating sliding motion using hemispherical riders was found[5] to be 0.45, a value lower than that obtained by Savage[3] using non-reciprocating motion and sliding graphite on copper. However, the value of 0.45 decreased with increasing pressures of such vapours as ethyl alcohol and heptane and atmospheres comprising steam, oxygen and air respectively. With water vapour, the friction coefficient was of the order of 0.18 at a pressure of 4 mm Hg. Presence of impurities in the graphite itself appeared to reduce the lubrication efficiency of these gases.

A fall in the coefficient of friction of graphite occurs if experiments are conducted at elevated temperatures. However, temperatures in the range of 800-1600°C are necessary before any appreciable change in the frictional behaviour is noted[6]. The mechanical strength of graphite is shown to remain constant irrespective of temperature. The reason for the lowered coefficient of friction is not clear.

24.2 *Carbides*

Mordike[7] has carried out friction experiments by sliding various carbides produced by powder metallurgy technique using light loads and low rubbing speeds. Measurements on couples of similar carbides showed a drop in the coefficient of friction at a temperature apparently characteristic of the material. Boron carbide was an exception which showed a steady increase in the coefficient of friction as the temperature in the experiment was increased. With tungsten carbide, the outgassed specimen at room temperature showed a coefficient of friction of 0.8, which fell to about 0.35 in the temperature region of 900°C. With titanium carbide, the room temperature value of μ was below 0.40 and this was nearly halved at around 900°C. In spite of the high values of friction, the wear rates of the carbides were low.

REFERENCES

1. Bryant P J, Gutshall P L and Taylor L H (Ed), *'Mechanisms of Solid Friction'*, Elsevier (1964).
2. Bryant P J, Taylor L H and Gutshall P L, *Trans 10th Nat. Vacuum Symposium*, (1963), pp. 21-26.
3. Savage R H, *Jl Appl Physics*, (1948), *19*, 1.
4. Savage R H and Schaefer D L, *Jl Appl Physics*, (1956), *27*, 136.
5. Rowe G W, *Wear*, (1960), *3*, 274.
6. Rowe G W, *Wear*, (1960), *3*, 454.
7. Mordike B L, *Wear*, (1960), *3*, 374.

CHAPTER 25

WEAR OF CAST IRON

The material cast iron is essentially an iron-carbon alloy with the total carbon content in the region of 3%. Part or all of the carbon appears as carbide particles and, when the carbon is free, flake or nodular graphite comprises a microconstituent depending on the casting technique employed. A very low silicon content is a prerequisite for producing white iron where the carbon exists as plates of carbides and a broad classification of cast iron according to its microstructure can be made as follows:

(1) graphite flakes distributed in a matrix of pearlite with or without some free ferrite;

(2) nodular graphite distributed in a pearlite matrix with some free ferrite;

(3) a mixture of cementite and austenite as in white iron;

(4) partially nodularised graphite in a ferritic or pearlitic matrix as obtained by heat treating white iron to produce malleable iron.

The mechanical strength of cast iron is influenced by its microstructure. For example, ceteris paribus, nodular graphite is associated with a higher tensile strength than the flake variety. Various additions of nickel, chromium, vanadium, titanium etc., are made to cast iron to control the tensile strength or the bulk hardness of the material. Not surprisingly, cast iron constitutes a popular material for many engineering applications. In situations involving wear, it has been used to resist abrasion as in shot blasting equipment and chutes to transfer minerals. It finds application in dry sliding situation such as automobile brake shoes and drums or plain bearings. Cast irons have found successful application in engine liners where the interaction is under lubricated reciprocating movement.

25.1 *The Role of Graphite*

The general model to explain wear of metals could, perhaps, be extended to cast iron provided it was recognised that the sliding surfaces possess a complex microstructure. The excellent wear resistance of grey iron under lubricated sliding condition has been attributed to the presence of graphite in the microstructure. It has now been well demonstrated, however, that the general view that the basal planes of graphite shear easily and provide a continuous source of a solid lubricant is not valid under all circumstances. It is important that the correct type of contaminant is present in the atmosphere for the high interlamellar energy of graphite to be lowered before it can be detached as particles from the matrix and then be smeared over the interface. If the graphite is detached from the cast irons, the result will be the generation of a series of micro voids and there is a view that these act as emergency reservoirs for oil which plays a beneficial role under boundary lubrication condition.

25.2 *Hardness*
There appears to be a correlation between the bulk hardness of cast iron and its wear propensity under abrasive condition. There are three[1, 2] categories of abrasive wear, viz.,

(a) high pressure abrasion such as it occurs in the moving and the static plates of a jaw crusher; one recommended material is an austenitic high manganese iron[3];

(b) high impact abrasion, e.g. iron shots used in cleaning of castings or ship's plates by blasting;

(c) sliding abrasion, as in pipes conveying abrasive materials such as foundry sands.

The criterion governing the service life of these components is believed to be a high bulk hardness. Both plain carbon and alloyed white irons with a hardness range of 400-900HV have been used successfully. The effect of alloying elements is to produce a martensitic structure which increases the bulk hardness of the material as in Ni-hard, a cast iron with about 5% Ni and 2% Cr. An addition of chromium by itself produces a structure of carbides in a matrix of ferrite. This is less hard than martensite but up to 30% chromium is added to the cast iron for use in brick making dies because of the excellent corrosion resistance of this material. The hardness of this iron is about 450 HV but irons with Vickers hardness of 900 can be produced by alloying with 12-18% Cr and 2-4% molybdenum.

Large scale field trials and accelerated laboratory wear tests have been conducted[4] for ploughshare materials which were plain and alloyed flake and nodular cast irons. The total carbon content varied from 2.9 to 3.7% with a maximum phosphorus content of 0.12%. The alloying elements were low. For example, nickel varied from 0.1 to 1.0%, chromium from 0.1 to 0.36% and the maximum molybdenum content was 0.24%. No hardness values are quoted but it is stated that there is a correlation between the Brinell hardness number and the wear resistance of irons cast in moulds of low thermal conductivity and the same is not true for chilled irons. The authors suggest that the reason for this was that, for the un-chilled irons, the structure comprised a uniformly distributed amount of graphite and pearlite. For the chilled irons, the carbides were distributed in an erratic manner in the pearlitic matrix. However, it is confirmed that abrasive soil resistance was the most superior when the iron was white, that is there was no free graphite in the microstructure.

Talanov and Chelushkin[5] suggest that the abrasive wear resistance of a heterogeneous material such as grey iron should be dictated by the algebraic sum of the hardnesses of the several microconstituents present in the matrix. Thus the relative abrasive wear resistance q_R of the bulk material can be equated with q_i, the relative wear resistance of the microconstituents $i = 1, 2, 3 \ldots\ldots n$, as

$$q_R = \sum_{i=1}^{i=n} V_i q_i \qquad (25.1)$$

where V_i = proportion of volume occupied by the constituent and

n = the number of microconstituents in the material

The authors[5] have conducted abrasive wear experiments under reciprocating sliding condition. The interface was filled with a lubricant containing 30% abrasive material consisting of equal parts of silica sand, corundum, iron filings and mill scale to simulate the environment in machine tool slideways. The specimens were cast and machined from a single cupola melt so that the chemical composition remained constant and the hardness of various experimental samples were varied by heat treatment. The experimental results suggested that there was a correlation between the relative wear resistance and the cumulative hardness of the non-graphite constituents. This was called the conditional hardness, H_c, such that,

$$H_c = \sum_{i=1}^{i=n} A_i H_i \tag{25.2}$$

where A_i is the area occupied by the microconstituent i and H_i is its microhardness. The microhardness of graphite is taken as zero, and as H_c increases, the relative wear resistance of the iron improves. It was demonstrated finally that, provided the specimens did not contain residual stresses or cracks, their wear resistance q could be expressed in the following form,

$$q = C \frac{H_c}{\log m} \tag{25.3}$$

where C is a constant and m is a measure of the number of graphite particles in the surface which is abraded. It would be interesting to verify equation 25.3, especially by incorporating various additives in the iron and measuring the microhardness of the constituents produced.

A high bulk hardness as measured by a Brinell or Vickers equipment is known to give rise to excessive wear during reciprocating sliding. It is stated[6] that hard surfaces, as produced by casting in chill moulds or subjecting components to flame hardening, wear readily if both members of a sliding couple are of similar hardness. The phenomenon is possibly associated with residual stresses due to a particular manufacturing technique or due to the presence of undercooled graphite as in the case of chill castings.

For abrasive conditions, measurements show[7] that residual stress has no effect on the rate of wear of cast iron. It is reported[8] that, even in a corrosive environment, the resistance to abrasion is good provided the bulk hardness of the material is high as in white irons.

25.3 *Lubricated Sliding Wear*

There does not appear to be any literature pertaining to the reciprocating sliding wear of white iron. However, a survey of some 30 machine tool slides has been made to obtain statistics on their wear resistance[9]. The main conclusions are that low phosphorus irons with a ferrite content of more than

10% wear badly. Microstructure, however, is unimportant provided the surfaces are well prepared, fully lubricated and kept free from grit and vibration. The role of surface finish in the behaviour of lubricated sliding surface has been investigated by controlled experiments[10]. Taking the load to seizure

Table 25.1 Surface Finish and Seizure Load of Cast Iron[10]

Surface Preparation	Finish μm	Seizure Load, kg
Machined	1.6	450
Ground	0.3	Did not seize at 1125 kg
Lapped	0.2	Did not seize at 1125 kg

as the criterion for wear resistance, it is shown (Table 25.1) how seizure occurs readily when rough surfaces are used in the experiment.

Cast iron cylinders are finished well before being put into service and observations[11] confirm that surface damage under boundary lubrication condition is minimised provided the amount of free ferrite does not exceed 10% and the phosporus content is increased.

Toresson and Olsson[12] have taken measurements on marine engine liners in service and carried out laboratory tests on representative materials using a pin-disc machine. In the latter case, the disc was continuously lubricated with oil which contained abrasives in the form of debris collected from engines in service. It was concluded that, to combat wear under these conditions, the graphite should be finely distributed but not in the undercooled interdendritic form. Grain boundary carbides as produced by small additions of vanadium, titanium or chromium improved wear resistance as did 0.1 to 0.3% copper.

Dumitrescu[13] examined the wear pattern of alloyed cast iron piston rings running for 2000 hours on a test engine. The rings were cast in greensand moulds and machined. Microstructural examination and hardness determination were carried out on the actual piston rings. The phosphorus content of the irons was of the order of 0.5% with copper in the range of 0.2 - 0.8%, chromium not greater than 0.86%, nickel below 0.9% and a vanadium addition of 0.18%. Not all the alloys were added simultaneously. Examination of the loss in weight of piston rings after 1000 hours' running (table 25.2) leads to the conclusion, contrary to the view expressed by other workers, that undercooled graphite results in a more wear resistant iron than when the graphite flakes are uniformly distributed.

Table 25.2 Weight Loss of Piston Rings[13]

Alloying Element	Type of Graphite	Average Weight Loss %
Copper	Uniformly distributed graphite flakes	6.48
Copper	Undercooled graphite	2.30
Chromium & Nickel	"	14.31
Chromium & Vanadium	"	8.33

The hardness of the last three irons in table 25.2 is also quoted to be higher than the iron alloyed with copper showing a random distribution of flake graphite. The induction hardened cylinder liners were made of grey iron alloyed with copper and wear on them was slight. Table 25.2 suggests that an iron with nickel and chromium or vanadium is not wear resistant. This seems unreasonable and the experiments should be repeated.

Lubricated wear studies would appear to need very careful control of the atmosphere. This is highlighted in the work of Takeuchi[14] who showed that wear of cast iron becomes excessive if the lubricant is heavily oxidised. Lack of control of the atmosphere could possibly account for the absence of any logical pattern in table 25.2. Incidentally, Takeuchi[14] shows that wear can be minimised by decreasing the spacing between the graphite flakes.

Regarding the shape of the graphite phase, there appears to be agreement in large measure that the nodular variety in a pearlitic matrix is more wear resistant than a corresponding flake graphite structure[15, 16]. Wear tests in a roller machine using both flake and nodular irons have been reported[17]. It was observed that the surface destruction of the specimens was the least when the matrix contained large nodules of the order of 50 microns in diameter. The role of microstructure in wear resistance was more positive for sliding contact of the iron on steel than when experiments were performed under rolling contact[18]. For the former situation, a nodular iron was only superior to one containing flake graphite when the matrix was pearlitic.
It was claimed that whatever the matrix, the nodular irons showed superior rolling wear resistance.

25.4 *Non-Lubricated Sliding Wear.*
Brake drums, discs and clutch plates undergo severe sliding wear in the absence of an external lubricant. From examination of commercial vehicles, two main types of failure of these components have been classified, viz.,

(1) Wearing and scoring.

(2) Heat-checking.

Angus, Lamb and Scholes[19] conclude that wearing and scoring can be avoided if the structure of the iron is mainly pearlitic with a bulk hardness of the order of 250 HB. It is necessary to control the amount of phosphorus as the resistance of the iron to thermal shock is reduced when this element exceeds 0.25%. The authors[19], possibly, infer mechanical failure rather than loss of metal by sliding but the work of Sulea[20] shows that the wear resistance of railway brake shoes doubles when the phosphorus content is increased from 0.3 to 0.8% and becomes six times more when the phosphorus content is increased to 1.5%. Nakai et al[21] have shown that an iron containing 0.7% phosphorus with additions of copper and chromium provides a wear resistant material in the brake shoe applications.

Heat checking can occur by a sudden phase transformation either by continuously running below the critical temperature of transformation or by sudden application of load which may result in local areas being raised above the critical temperature[19]. When the contacting surfaces remain continuously at about 760°C, which is a little below the critical temperature, as it may occur on brake drums of vehicles moving in peak traffic condition in a city, the graphite particles get oxidised from the surface and the pearlite lamellae are spheroidised. This causes a reduction in the bulk hardness of the surface and, hence, cracking by heat-checking. Whereas the problem occurring under continuous temperature condition can be minimised by providing a structure capable of dissipating heat rapidly, for example by increasing the total amount of graphite and thus the thermal conductivity of the material, heat-checking due to sudden application of load inevitably establishes a steep temperature gradient in the body of the component. The net effect is that up to 0.5 mm of the material from the surface becomes martensitic because of the heating followed by quenching in air of the iron surface. This results in a differential volume expansion between the surface layers and the bulk of the component, giving rise to cracking. It has been recorded that the sudden temperature rise can be decisively above 900°C under this condition and, in extreme cases, localised melting may occur. It would appear that a high graphite content minimises heat checking of cast iron because of the following considerations[19].

(1) The thermal conductivity of the brake drum increases directly with the amount of graphite in the cast iron.

(2) An excess of graphite means extra lubricating surfaces which would diminish the cold welding propensity of the metallic asperities.

(3) The higher the graphite content of the iron, the lower is the modulus of elasticity of the material and hence a reduction in the instantaneous stress that can develop during sudden heating and cooling.

The authors[19] show that very finely divided undercooled graphite lowers the wear resistance and makes the material more susceptible to thermal shock. It is suggested that the associated ferrite with an undercooled graphite colony provides the soft constituent to be torn out of the matrix relatively easily. Heat-checking occurs probably because the graphite goes into solution in the iron readily and the austenite, thus formed, will transform to martensite upon cooling, resulting in a differential volume expansion and hence cracking.

The effect of microstructure and load on the wear rate of cast iron has been investigated under dry sliding condition using a pin-disc machine[22]. The pin was a vertical cylinder 0.635 cm diameter x 5 cm long and the disc, 7.62 cm diameter, was a heat treated steel of hardness 320 HV. The experiment was carried out in room atmosphere at varying loads up to 6 kg. The pins were polished on emery papers and four types of pearlitic flake graphite irons with ferrite contents of 5, 10, 15 and 20% respectively were tested. The distribution of graphite was random in some pins whereas in others the graphite was interdentritic. The main conclusions from these experiments were that the amount of wear increased in a linear manner as the applied load was increased until at the transition load when the loss of pin weight was excessive (Fig. 25.1). By examining the debris, the authors[22] show that, below the transition load, the nature of wear is largely metallic. That is, the mechanism is the welding together of the metallic asperities by plastic flow followed by shearing of these junctions. At heavy loads, frictional heat causes oxidation and the wear process is by oxide film removal as well. Regarding the microstructure, it is shown that the best wear resistant materials are those having the graphite distributed in a random manner.

Similar experiments[23] with spheroidal graphite cast iron show that the accicular bainitic microstructure provides the best wear resistance under dry unidirectional sliding motion. For iron pins, a hard white layer forms at the sliding interface and contains oxidised cracks. A nodular iron has been shown to give a lower wear rate than a flake graphite iron below the transition load[24].

25.5 *Concluding Remarks*

The bulk of the published information on the wear of cast iron would appear to be from observations on the actual application of the material in machine tools, automobiles, locomotives and the agricultural field. There have been some laboratory investigations but the overall impression from these experiments or from observations on practical situations is that certain conclusions are anomalous, for example the role of phosphorus or of bulk hardness in the

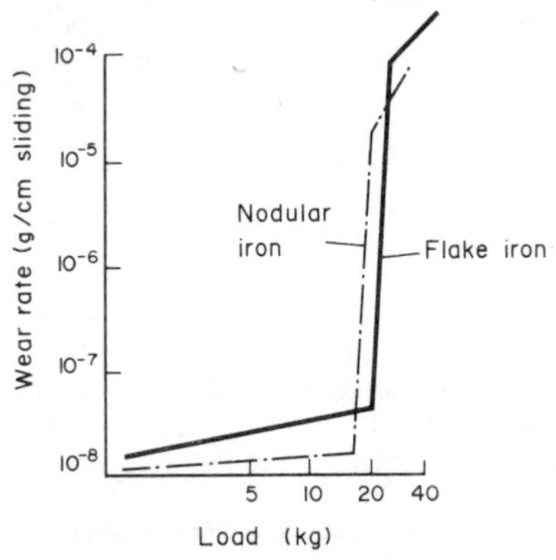

Fig. 25.1 Variation of wear rate with load of flake and nodular cast irons showing the transitional nature of wear (Eyre, Iles and Gasson[24]).

wear rate of cast iron. This is probably because of the complex microstructure of the cast iron and the sensitivity of the graphite phase to the surrounding atmosphere from the point of view of interfacial interactions. Disregarding the anomalies, however, the majority opinion would appear to be that, for abrasive use, a high bulk hardness with an absence of residual stress is necessary and a nodular graphite structure in a pearlitic matrix is desirable. For dry sliding wear, the following characteristics are beneficial if a high resistance to wear is desired:

(1) The iron should have a full pearlitic matrix with a hardness of about 250 HB. This is much below what is recommended for abrasive wear application employing white iron.

(2) There should be an upper limit of 0.25% on the phosphorus content and the amount of graphite should be high.

(3) Undercooled graphite and free ferrite should be absent.

(4) SG irons are superior to flake graphite cast irons, especially if the structure is bainitic.

Further laboratory investigations on the role of microstructure and mechanical properties in the wear mode of cast irons should be useful. The effect of additives in the iron should receive special attention.

REFERENCES

1. Angus H T, *'Physical and Engineering Properties of Cast Iron'* (1960), BCIRA.
2. Angus H T, *J1 BCIRA*, (1962), *10*, 80.
3. Lyadskii V B and Shabalov V I, *Russian Castings Production*, (1964), p.561.
4. Mohsenin N, Womochel H L, Harvey D J and Carleton W M, *Agricultural Engineering*, (1956), *37*, 816.
5. Talanov P I and Chelushkin A S, *Russian Castings Production*, (1964), p.127.
6. Brewer R C, *Engineering*, (1961), *192*, 84.
7. Shapkin V M, *Russian Castings Production*, (1965), p.72.
8. Deardon J and Swindale J D, *J1 Iron and Steel Inst*, (1957), *185*, 227.
9. Angus H T, *Wear*, (1957), *1*, 40.
10. Moore W H, *Product Engineering*, (1958), *29*, 63.
11. Tommis N, *Hepolite Bulletin*, (1964), *19*, 10.
12. Toresson S and Olsson B, *BCIRA Conference on Engineering Properties and Uses of Iron Castings*, (1956), 28-30th November, London.
13. Dumitrescu T, *Revue Roumaine De Metallurgie*, (1961), *6*, 47.
14. Takeuchi E, *Wear*, (1970), *15*, 201.
15. Lyadskii V B, *Metal Science and Heat-Treatment*, (1963), Nos 11-12, Nov-Dec, p.656.
16. Serpik N M and Kantor M M, Ibid, (1964), Nos 7-8, Jul-Aug, p.451.
17. Vashchenko K I and Zhuk V Ya, *Russian Castings Production*, (1961), p.496.
18. Stahli G, *Giesserei*, (1965), *52*, 406.
19. Angus H T, Lamb A D and Scholes J P, *J1 BCIRA*, (1966), *14*, 371.
20. Sulea P, *Metallurgical Researches, Paper No. 8, Conference at Bucharest*, (1964), Jan 12-15, pp 79-93.
21. Nakai M, Saito S and Okabayashi K, *Imono*, (1961), *33*, 24.
22. Hodgson M J and Eyre T S, *Jl BCIRA*, (1967), *15*, 257.
23. Wilson F and Eyre T S, *Wear*, (1969), *14*, 107.
24. Eyre T S, Iles R F and Gasson D W, *Wear*, (1969), *13*, 229.

CHAPTER 26

WEAR OF ALUMINIUM-SILICON ALLOYS

The metal aluminium is light, has good electrical and thermal conductivity and its resistance to corrosion is used with advantage. Pure aluminium itself has poor foundry characteristics but suitable alloying elements can overcome this and the alloys of the metal find application in many fields either in the as-cast or heat treated condition, a typical area being the internal combustion engines. Aluminium alloys, because of their lightness, began to replace the cast iron pistons around 1920. Some typical aluminium alloys employed in tribological situations are shown in Table 26.1. The Y alloy is known to possess high hot strength and finds application where the thermal load is severe such as in diesel engines. Its wear resistance is poor and it has a high coefficient of thermal expansion. The alloy LM12 has been used for decades but it appears to have very inferior wear resistance. Lo-Ex finds extensive application in petrol engines and a hypereutectic alloy with a silicon content of the order of 20% has been used frequently in Europe and the USA but it is still in an exploratory stage in the UK.

Table 26.1 Composition of a few Aluminium-Silicon Alloys

Alloy	Cu	Ni	Mg	Fe	Mn	Zn	Pb	Sn	Si	Al
Y	4	2	1.5	-	-	-	-	-	-	Remainder
LM12	10	0.5	0.3	1.5	0.6	-	-	-	2	Remainder
Lo-Ex	0.5-1.3	2-3	0.8-1.5	0.8 max	0.5 max	0.1 max	0.1 max	0.1 max	11-13	Remainder

26.1 *Effect of Silicon on Wear*

An important constituent in these alloys is the element silicon and, accordingly, attempts have been made to find a correlation between the rate of wear and the amount of silicon. Vandelli[1] has carried out laboratory experiments to study the rates of wear of aluminium-silicon alloys sliding under reciprocating condition on cast iron. He concludes that resistance to wear is probably governed by the way the silicon particles are distributed and not on the amount of the constituent in the alloy, a fine dispersion of the silicon needles giving a high resistance to wear.

Okabayashi et al[2-6] have studied the friction and wear of a few commercial alloys under controlled condition. Various alloys, usually heat treated, with silicon contents of 0.18, 6.20, 11.22 and 21.60 respectively were slid against mild steel and grey iron discs. Scuffing occurred when similar alloys were in rubbing contact at high speeds of sliding[2]. The hypereutectic alloy showed little scuffing and the rate of wear was low. It was shown that as the coefficient of friction increased, there was a rise in the rate of wear[3].

In all combinations, after a running-in period, the weight loss was linear with sliding distance. Generally, the rate of wear of a pin was higher than the ring and the authors attribute this to the higher temperature of a pin developed during sliding[5]. The conclusion about silicon was that the total quantity of silicon rather than the nature of the distribution of the particles in the matrix determined the rate of wear, a high silicon content giving superior wear resistance.

A study of the metallurgy of aluminium-silicon alloys shows that, like cast iron, the material possesses a complex microstructure. A correlation between composition or microstructure and the friction and wear mode of these alloys is not decisive at the moment. However, it is experienced[7] that the heat treated hypereutectic alloys give favourable properties from the view point of wear resistance when used in such applications as clutches, cylinder liners and pistons for automobiles.

The full potential of these alloys is realised by solution treating the cast or wrought components and then aging at an optimum temperature. A clue to the phenomenology of wear of these alloys can be obtained by metallurgical examination of surfaces and subsurfaces of components which have been slid under load and some observations from laboratory experiments are discussed in what follows.

26.2 *Deformation of a Bush*

A hypo eutectic alloy bush aged at 170°C was run against steel and cast iron pins at two different loads for comparable sliding distance. Figure 26.1 shows that wear along the circumference of the bush is not uniform. It is relatively easy to establish by metallographic techniques such as looking at transverse sections of the track that the interface becomes crumpled at the early stages of running the machine due to a ploughing action of the pin. That is to say that, with the onset of a tangential pull, an amount of material is quickly removed by ploughing but the wear track of the bush flows plastically ahead of the pin and work hardens (Fig. 26.2). During the next encounter, the pin probably slides over this hump and removes an amount of metal by abrasion. At heavy loads, the steel pin continues to remove metal and the groove is always deeper than the original surface of the bush but at light loads, the track shows raised lumps of metal which are higher than the bush surface. This suggests that the work hardened hump ahead of the pin fractures and is removed readily when the load is heavy. At light loads, once the surface has work hardened to a depth, an equilibrium situation is reached and wear occurs by abrasion and spalling of small surface layers. Metallographic observation at light loads shows a hard deformed layer below the worn track in the body of the bush. By contrast, this is absent when the load is heavy confirming the supposition that as the subsurface deforms it fractures and is removed continuously.

Gross plastic flow occurs on either side of the wear track and the work hardened layer follows a similar flow path to that of the shear plane developed during metal cutting. As sliding proceeds, a crack develops from the surface and follows the interface between the deformed layer and the bulk material and wear particles are produced this way. Cracks also initiate in the deformed layer and propagate through it to produce wear debris. The total amount of wear is low with cast iron pins possibly because of a diluting effect by the graphite flakes.

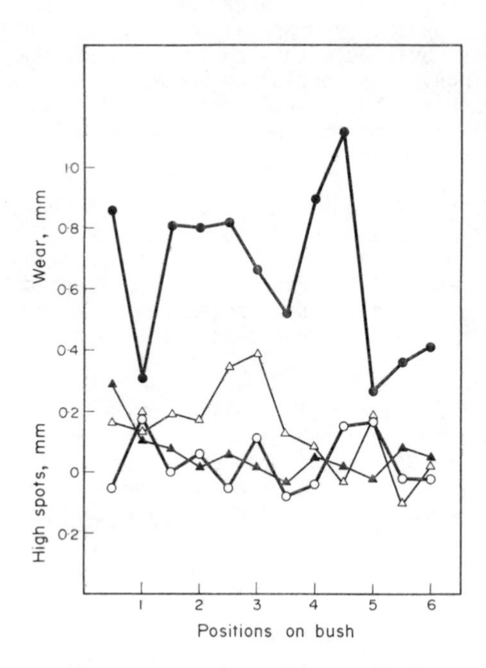

Fig. 26.1 Final wear on a 10 cm diameter hypoeutectic aluminium - silicon alloy bush measured after sliding distances (s) as shown below. o steel pin at 1.6 kg; s = 580 788 cm; ● steel pin at 5.6 kg; s = 580 444 cm; △ iron pin at 1.6 kg; s = 703 482 cm; ▲ iron pin at 5.6 kg;s = 895 986 cm.

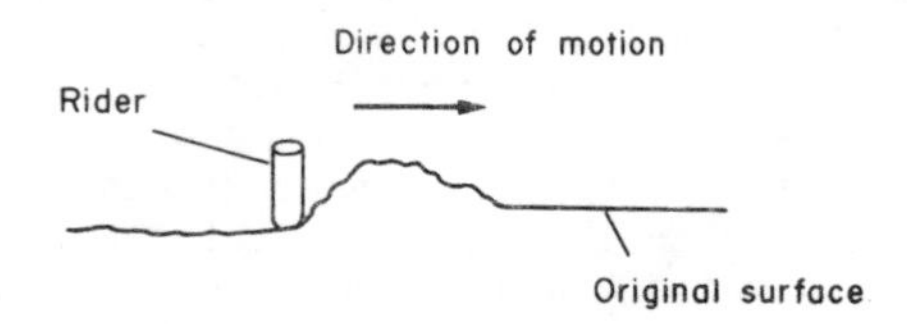

Fig. 26.2 Plastic flow and piling up of metal ahead of a rider (schematic).

26.3 *A Physical Model*

Similar observations on the surface and the subsurface of aluminium-silicon alloy pins suggest a physical model[8] of wear as shown schematically in Fig. 26.3, which shows a model surface and subsurface of three conical asperities before and after sliding. The asperities are shown to be covered further with micro-asperities. With the onset of a tangential pull of the loaded pin, the asperities flatten in the direction of motion and plastic flow occurs below the original surface. With progressive encounter, the asperities flatten further and the zone of the deformed subsurface extends. A microscopically rough surface is thus created meaning that there are large areas which are never in direct contact with the opposing member of the couple. As the traversals continue, the layered surface and a zone below it will work harden to a limiting value but further micro-microasperities will be created upon the original conical protuberances when they are still in a plastic condition. This is suggested because a transverse section of a wear track shows these small peaks at high magnification and the situation is analogous, for example, to moderately hydrated clay upon which a glass slide has been pressed and then lifted when the clay surface becomes corrugated. The situation thus is this.

Junctions form due to an application of load and these grow as a result of the tangential interaction at the interface. As long as the surface is in a plastic condition, the metal from the aluminium-silicon alloy pin is initially removed in the form of transferred fragments onto the cast iron or steel bush. With continued encounter, the surfaces work harden but plasticity of the newly created micro-microasperities is maintained by thermal softening since the true contact pressure is very high there.

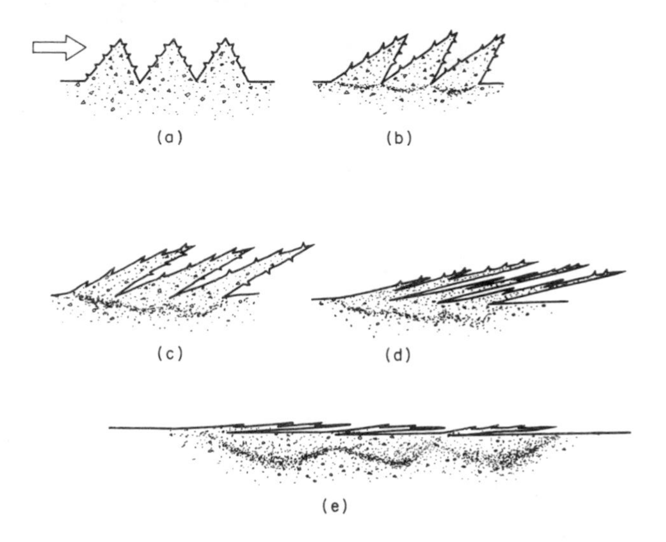

Fig. 26.3 A physical model showing flattening of asperities. Increasing sliding distance a→b→c→d→e. Microasperities are present before sliding but micro-micro-asperities are created during sliding. There are areas of the asperities which are never in contact with the opposite member of the couple. A subsurface deformed layer forms. (Sarkar[8]).

However, it is suggested[8] that the predominant force in producing wear debris is due to an elastic encounter. Examination of the wear scar using the technique of scanning electron microscopy shows that wear particles are produced as small spalled fragments from the surface. Spalling of small areas is attributed to fatigue and, although contact occurs at the high micro-asperities which generate plastic junctions the maximum Hertzian stress is below in the work hardened subsurface. This region of maximum stress is under repeated stress cycling, being zero when the high spots surrounding it are not in contact with the bush as it must happen on an uneven surface like this. When the critical number of cycles has been exceeded, a fatigue crack initiates and propagates to give a wear fragment.

26.4 *Wear Rate*
Since the surfaces are under a mixed mode of plastic and largely elastic

interaction, linearity of the wear rate with load should not be expected. This was seen[8] to be true in experiments where both the aging temperature and the amount of silicon were varied. A typical example is shown in Fig. 26.4 where the rate of wear, q, of a hypoeutectic aluminium-silicon alloy running on grey iron is plotted against load, W. The first regime of wear is curvilinear and this gives way to a second regime at a load of about 3 kg on a 6.25 mm diameter pin. The pins beyond this load, however, show gross plastic flow and this should not be considered as wear but as failure of the bulk material.

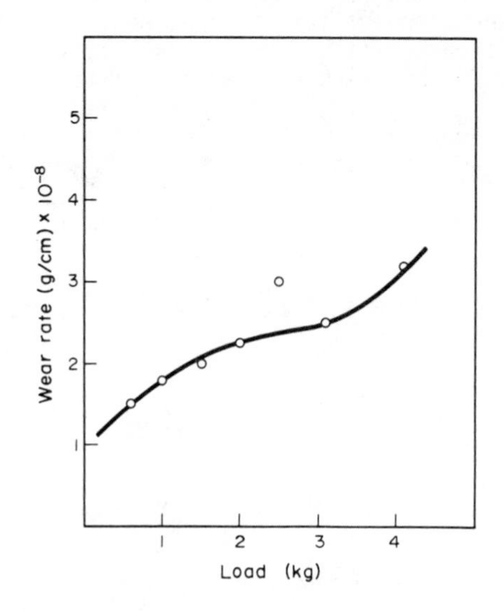

Fig. 26.4 Wear rate of hypoeutectic aluminium-silicon alloy pins with load. Pins aged at 200°C for 5 h giving hardness of 137 HV10. Speed 196 cm/s. Grey iron bush (Sarkar[8]).

The experiments of the author[8] led to the following conclusions:

(1) Up to the load where failure of the bulk material occurs, the rate of wear, q (g/cm), can be related to the load W (g) as

$$q = K\,W^{\alpha} \qquad (26.1)$$

where K is a constant which is not dimensionless and will vary in magnitude according to the amount of silicon in the alloy or its thermal treatment prior to sliding; α is a constant and probably has a value between 0.40 and 0.60.

(2) Wear is accentuated if the aluminium alloy slides on itself.

(3) Overaging gives a higher wear rate and an underaged alloy gives a lower wear rate than the alloys which are aged in the optimum temperature range.

(4) The hypereutectic alloy wears more than the hypoeutectic material.

REFERENCES

1. Vandelli G, *Alluminio and Nuova Metallurgia*, (1968), *37*, 121.
2. Okabayashi K, Nakatani Y, Notani H and Kawamoto M, *Keikinzoku (Light Metals Tokyo)*, (1964), *14*, 57.
3. Okabayashi K, Nakatani Y, Notani H and Kawamoto M, *Ibid*, (1964), *14*, 71.
4. Okabayashi K, Kawamoto M and Notani H, *Ibid*, (1966), *16*, 38.
5. Okabayashi K, Kawamoto M and Notani H, *Bulletin of University of Osaka Prefecture A*, (1966), *15*, 153.
6. Okabayashi K and Kawamoto M, *Ibid*, (1968), *17*, 199.
7. Stonebrook E E, *Modern Castings*, (1960), *38*, 111.
8. Sarkar A D, *Wear*, (1975), *31*, 331.

CHAPTER 27

DESIGN FOR WEAR RESISTANCE

The process of wear may be defined as loss of material from the interface of two bodies when subjected to relative motion under load. The bulk of engineering systems involves relative motion between components fabricated from metals and non-metals and three main types of wear have been identified as follows:

(1) adhesive wear;

(2) abrasive wear;

(3) fretting.

These have been discussed in detail in the preceding chapters and the importance of wear and the principles employed to minimise it are briefly discussed here. However, prior to this, it is necessary to outline the mechanism of three other forms of wear which are included in the classification of wear processes in metals and materials.

These are

(1) Fatigue wear.

(2) Erosive wear.

(3) Cavitation Erosion.

27.1 *Fatigue Wear*

It is probable that the predominant mode of most types of wear is by spalling of material from the interface by fatigue whether the nature of movement is unidirectional or reciprocating. To classify a particular type of failure as fatigue wear may, therefore, be confusing. However, for the sake of classification, fatigue wear may be reserved to identify the failure of lubricated contacts in such situations as ball and roller bearings, gears, cams and friction drives[1]. Material loss is in the form of spalling of surface layers and pitting as in gears.

The fatigue cracks are again believed to start below the surface at a point where the shear stress is maximum. Obviously an improvement in the life of these elements can be achieved by working at low contact load and the most popular method in industry is to produce components with an optimum depth of a hard case coupled with a high degree of surface finish. The purpose of a hard outer case as provided by carburising or nitriding is to provide a mass with a high endurance limit in a region most vulnerable to crack initiation.

27.2 *Erosive Wear*

Erosive wear has been defined as the process of metal removal due to impingement of solid particles on a surface. Erosive wear is deliberate as in the case of cleaning of castings or ships hulls by shot blasting, but destructive loss of expensive material occurs in such situations as gas turbine blades or refractories in electric arc melting furnaces.

The degree of wear has a relationship with the impingement angle of the particle relative to the recipient surface (Fig. 27.1). Ductile materials appear to deform and possibly work harden when struck normally but, at a critical angle of about 20^0, metal is removed by a cutting action. The brittle materials fail by cracking of surfaces when the force of impact is normal. It follows that an initially ductile component will eventually work harden and fail in a brittle manner.

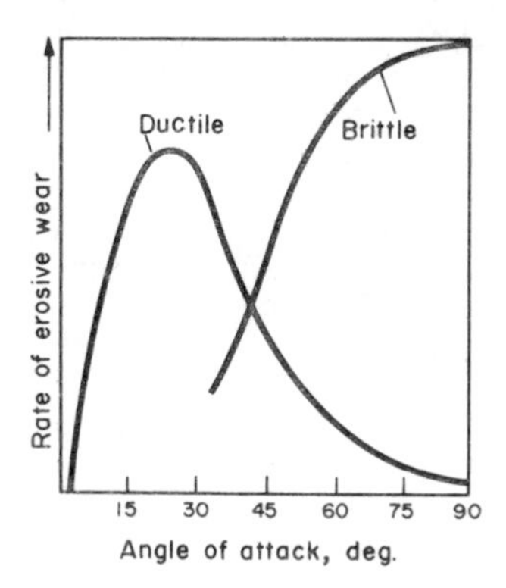

Fig. 27.1 Dependence of erosive wear rate of ductile and brittle solids on the angle of attack of impinging particles. (Wright[1]).

Attempts are made to manipulate the impingement angle by modifying the design of the component itself or by protecting the surfaces with a material such as rubber. Controlling the angle of impingement in an industrial situation is not easy and it is pertinent to speculate whether surfaces should be coated with media which can be replenished easily as a routine maintenance procedure.

27.3 *Cavitation Erosion*

Cavitation erosion occurs when a solid undergoes movement at high velocities in a liquid medium, a typical case being ships propellers. Whereas erosive wear implies a purely mechanical action, cavitation erosion is believed to be tied up with the formation of bubbles within the liquid medium through which the solid component is shearing. Cavitation erosion occurs also in lubricated bearings.

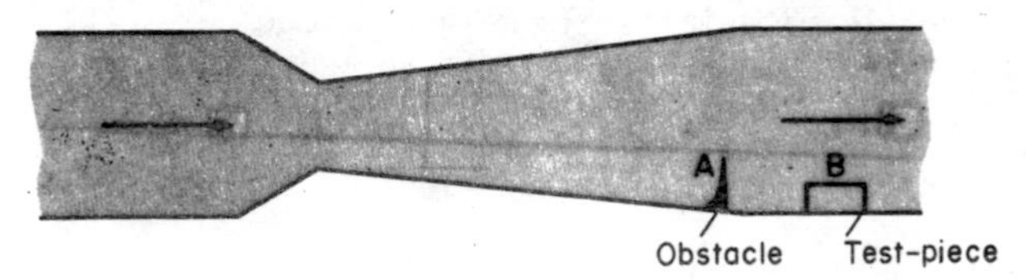

Fig. 27.2 A venturi tube for cavitation erosion experiments. A, obstruction; B test piece under study.

For laboratory testing of materials, the most preferred[2] method is the use of a venturi tube (Fig. 27.2) where the fluid at high velocity encounters an obstacle A. This creates bubbles which are allowed to impinge on the test piece further downstream.

At the throat of the venturi tube, the area of cross-section diminishes which has the effect of increasing the fluid velocity in that region and, according to Bernoulli's theorem, there will be a drop in pressure there. If the flow rate is high enough, this pressure drop may be lower than the vapour pressure of the surrounding liquid, in which case cavitation, that is formation of vapour bubbles, will be facilitated. As the bubbles are carried downstream they collapse in the parallel part of the tube where the pressure is higher than that at the throat. The collapse of a bubble gives rise to an instantaneous high local pressure. Cavitation erosion is attributed to the mechanical disintegration of the metal under these highly localised pressure pulses. The impact force thus produced has been shown to exceed the yield point of most metals.

Perhaps the most effective metals are those which are hard with a high yield strength or which work harden in service. The most suitable material is stellite and 18/8 stainless steel. Cast manganese bronze is a popular material for marine propellers. It has inferior cavitation erosion resistance to stellite or stainless steel but is resistant to marine corrosion and it has been demonstrated that corrosion aggravates cavitation erosion. Apart from hard facings of chromium to a suitable metal substrate, a resilient layer which can absorb the high impact energies is often used. Examples of the latter are polyurethene and neoprene, among others.

27.4 *Design for Adhesive and Abrasive Wear Resistance*

Both adhesive and abrasive wear can be regarded as the most common form of wear in engineering. Whenever two surfaces meet, there is the probability of adhesion at favourable points at the interface followed, possibly, by wear. It is unlikely that the interface will be free from grits and wear debris so that the mechanism of wear in such situations as plain bearings and cylinder liners is probably a combination of adhesion and abrasion.

To design against wear, it is undesirable that a couple be selected which comprise metals with mutual solubility. Thus a steel shaft running on a steel bearing is not a logical proposition although cast iron has been slid on itself successfully possibly because of the presence of graphite. Both the abrasive and adhesive wear laws show that the harder a component is the more resistant it is to wear and this is confirmed in practice. A difficulty with excessively hard materials is that they are liable to brittle failure and, to avoid mechanical failure, the component must be tough which is the characteristic of soft ductile materials. To maintain toughness, therefore, it would appear that hardness must be sacrificed. However, a tough core with a hard surface can be achieved by heat and surface treatment of components. Generally, in the absence of oscillatory motion, corrosion and high temperature, the following principles are applied when selecting steels:

(a) Low impact load: use can be made of hard carbides in the microstructure or the steel can be nitrided or carburised. Martensitic irons and steels can also be used.

(b) High impact load: The material philosophy is to use austenitic, stainless and Hadfield manganese steel.

Three broad classifications[3] of tribological practice have been made to minimise wear, viz., the use of a protective layer and the principles of conversion and diversion.

27.4.1 *Protective Layer.* A protective layer of oxide or sorbed gases inevitably forms on all surfaces and without this many dry sliding situations such as railway tracks would not have survived to give an acceptable life. Lubricants either in the form of a liquid or solid are invariably used in most situations. Lubrication has been briefly touched upon in discussing plain bearings in chapter 22 and will not be discussed further here.

Currently, there is a renewed interest in assessing various types of surface treatments prior to putting components into service with a view to achieving wear resistance. Broadly speaking, the treatment can be classified into two types, viz.,

(a) Deposition of metal or non-metal on the component.

(b) Diffusion treatment such as carburising or nitriding.

Examples of deposition are phosphates on cast irons and steel or tin on aluminium[4]. These have limited lives and protect surfaces from severe damage during running-in. Other forms of diffusion treatment are sulphiding, chromising and electrodeposited coatings. Wear resistant coatings are also imparted by metallising and spraying of metal or by hard facing techniques such as using welded overlays. There are detailed accounts of materials and processes used for such applications in many metallurgical text books.

27.4.2 *Principle of Conversion.* The principle of conversion allows wear of one part of the system to offer protection to more important components. An example is the use of cast iron piston rings which, if allowed to wear rapidly, should prevent scuffing of the cylinder liner.

27.4.3 *Principle of Diversion.* The less costly component, for example a plain bearing, is designed to wear to protect the more expensive journal from surface damage and wear. Of the journal bearings, white metals containing tin, lead, copper and antimony provide a soft sliding surface. However, if the load and speed are increased, harder bearing surfaces as provided by the bronzes must be employed.

27.5 *Importance of Wear*

Wear of parts means replacement and this by itself is expensive. Moreover, worn surfaces cause a loss of precision with the resultant sacrifice in efficiency. Every attempt should be made, therefore, to design for a minimum amount of wear.

There is sufficient evidence to show that at a characteristic load, depending on the properties of the material, a rapid transition occurs leading to severe wear from the preceding regime of mild wear (Fig. 25.1). It is imperative that moving parts of machinery must be so designed that the contact load does not exceed what is dictated by the regime of mild wear. It is not more expensive in the long run to employ materials which would possess high flow stress to allow for heavy loads in service.

Whereas every attempt must be made to minimise steady state wear, surface layers must be removed during running-in of new machines. The process of running-in probably involves the establishment of a subcutaneous deformed layer which will be hard enough to resist steady-state wear but surface peaks must deform and wear in order to inflict conformity to the interacting couple. The fact that extra mechanical work is done in deformation and wear at this stage can be seen in Fig. 27.3 which shows how the fuel consumption is high during the running-in process of a new automobile. Prediction of wear rate, however, is difficult because it depends not only on the load and surface speed but on other factors which are not always easy to avoid. There is always the presence of abrasives which may be extraneous or the debris which is produced by interaction and abrasives are known to increase the rate of wear. A high surface speed is beneficial because it gives a higher interfacial temperature which facilitates the formation of oxide films. In the case of petrol engines, about a gallon of water is produced for every gallon of fuel burned. The products of combustion contain oxides of carbon, sulphur and small amounts of bromine and chlorine compounds. If the engine wall temperature is low, the water and the products of combustion form acids corrosive to the cylinder wall and piston rings. The wear is accentuated as shown in Fig. 27.4. It is not surprising that a single wear law has not been

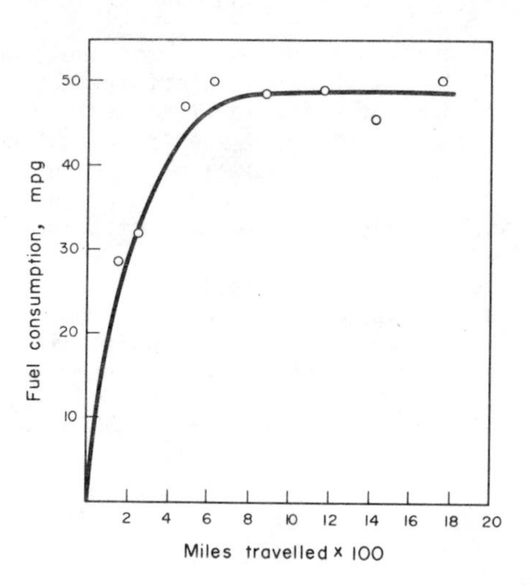

Fig. 27.3 Fuel consumption of a new automobile with sliding distance. The high fuel consumption during the running-in stage is obvious.

propounded which could be used with confidence to predict wear rates of interacting components. An important reason is that, in common with many technological systems, most tribological situations are compelled to perform under the combined action of several variables. For example, abrasion may be superimposed on a situation which was primarily designed for adhesive wear and the problem may be further aggravated by the system having to operate in a hostile environment. It is imperative, therefore, that to design against wear, the service conditions of the components must be identified. To predict wear rates, however, there should be continued investment in research with a view to obtaining information about the mechanisms of the various types of wear of interacting surfaces undergoing relative motion.

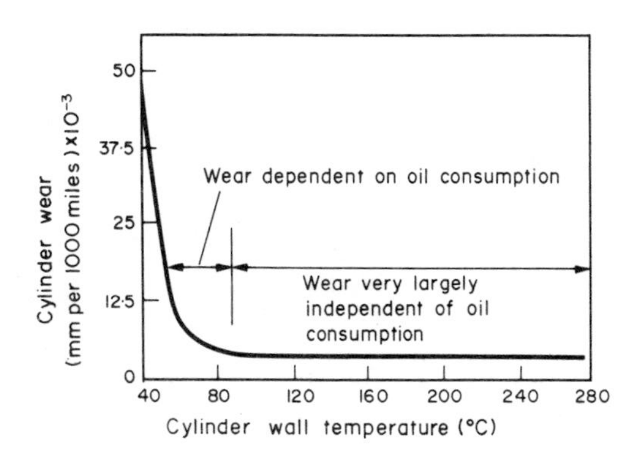

Fig. 27.4 Wear of an engine as a function of cylinder wall temperature.

REFERENCES

1. Wright K H R, *Tribology*, (1969), *2*, 152.
2. Beeching, R, *Product Engineering*, (1948), 110.
3. Lipson C, *'Wear Considerations in Design'*, Prentice-Hall, (1967).
4. Wilson R W, *DeIngenieur*, (1967), *79*, 100.

INDEX